Five Friends Discuss Global Warming

How It Started, Why It Won't Stop

W. Clark Dean

Copyright © 2025 W. Clark Dean

All rights reserved.

No part of this book may be used or reproduced in any manner whatsoever without written permission of the publisher, except in the case of brief quotation embodied in critical articles or reviews.

ISBN 978-1-964143-07-1

 Suncoast Digital Press, Inc
Sarasota, Florida

Printed in the United States of America

Cover illustration and all drawings in this book are by the author W. Clark Dean

CONTENTS

Introduction v

Part 1
We've Got A Problem!. 1

1. Those Darn Dinosaurs 3
2. The Conventional Wisdom 7
3. The Dinosaur/Gravity Problem. 11
4. Gravity and Centrifugal Force. 17
5. Pangaea 23
6. Electricity 27
7. The New Unconventional Wisdom 33

Part 2
The Rise & Fall Of The Dinosaurs 37

8. A Solar System Is Born 39
9. The Dimensions of the Early Solar System. 45
10. The Evolution of Life. 55
11. The Extinction of the Dinosaurs 67
12.. Continental Drift 77
13. Mountain Building. 83

Part 3
A New World Begins 93

14. The Survivors Evolve 95
15. 5An Omen and the Golden Age of Saturn 103
16. The Eye in the Sky. 107
17. The Land Before Time 117

Part 4
Catastrophe! 119

18. Gone in a Flash 121
19. Telltale Features of Mars and Venus 137
20. The Storytellers 147
21. Early Stories and the Ages of Earth 161
22. The Age of Uranus (The First Age) 169
23. The Age of the Moon (The Second Age). 171
24. The Deluge and The Age of Saturn (The Third Age) 175
25. The Age of Mercury (The Fourth Age). 191
26. The Age of Jupiter (The Fifth Age). 195
27. Exodus & The Age of Venus (The Sixth Age) 211
28. The Age of Mars (The Seventh Age). 215

Part 5
Calm and Understanding 219

29. The Present-Day Solar System 221
30. Climate Change 231
31. Closing Arguments 237
32. A Brief Timeline of the World 245

Epilogue 255
Acknowledgments 257
About the Author 259

INTRODUCTION

A Conversation Among Friends

Five friends—an astrophysicist, a climatologist, a geophysicist, a paleontologist, and an engineer—are walking down the street engaged in animated conversation.

The astrophysicist says, "My field of study is fascinating with its Big Bang, black holes, dark matter, gravity waves, a nuclear-powered Sun, and other really neat stuff. Together, they provide a theoretical explanation of how the observed universe works. Unfortunately, we sometimes have to resort to unconventional math to explain some of our theories. I'm particularly concerned by the lack of any hard evidence of dark matter that theoretically must exist as we try to explain the rotation of galaxies and the stars within our own galaxy. There are well-educated people advocating for an 'electric universe' where electricity is just as important as gravity. I'm looking into that as a possible answer."

The climatologist says, "Closer to home, finding a solution to the upsurge in global warming and rising sea levels has been dominating my activities over the last several years. We are trying to make sense of Antarctic ice core and other data showing that there was an Ice Age covering much of North America, northern Europe, and Asia 26,500 years ago profoundly affecting Earth's climate by causing drought, desertification, and a large drop in sea levels. A warming trend started in the Northern Hemisphere approximately 20,000 years ago and in Antarctica approximately 14,500 years ago, consistent with evidence for an abrupt rise in the sea level at the same time. This shows that our present-day global warming trend is not a new phenomenon, but has been going on since about the time of Noah's flood. As of this moment I have no idea how to stop its effects."

The geophysicist responds, saying, "I really enjoy my field of study with plate tectonics building mountains and moving continents around, and recently the discovery of what may be an asteroid impact crater in Mexico's Yucatán Peninsula that provides a smoking gun for the cause of the dinosaur extinction."

The paleontologist joins in, saying, "Speaking of the dinosaurs' extinction 65 million years ago, we recently discovered an unusual clue that may help uncover what caused these massive creatures' demise. Namely, no land animal having an adult weight of more than 50 pounds survived the event. Mammals, land turtles, snakes, lizards, birds and crocodiles that met this weight criterion survived and continued to diversify, but only the small survived. To add to the intrigue, my field of study has recently made a big discovery." After hesitating a moment he adds, "I mean a really big discovery. We found fossilized dinosaur bones in Patagonia that belonged to a creature that must have weighed 100 tons and stood 50 feet tall! It appears to be a close cousin of one found in Argentina back in 1987 that was of the same incredible size."

The engineer (the only non-PhD in the group) stops walking, ponders for a moment, and says, "How big did you say that dinosaur was?"

"100 tons and 50 feet tall," the paleontologist replies.

All five stop and look at each other. "That's really big!" they all agree.

Then, after a brief moment of glazed-eyed mental calculations, the engineer says, "Too big! That dinosaur is too big to stand, and with all deference to your dinosaur extinction explanation, my dear paleontologist friend, the real unanswered problem here is not how those huge dinosaurs became extinct, but how could they have existed here on Earth in the first place?" After a pause he adds, "And whatever is causing our present-day record-setting high temperatures may somehow be involved as well. Let's meet for lunch in the park next week. If the five of us put our heads together we should be able to work this out. This changes everything!"

Introduction

A few days later, after much soul-searching and head-scratching, the five friends meet for lunch in the park.

Astrophysicist: "I'm having difficulty explaining several tenets of my field such as how a mathematical singularity such as a black hole can exist in the physical universe, and why can't we find any dark matter. My hero, Albert Einstein once said, 'If at first, an idea is not absurd, there's no hope for it.' We need to find an absurd solution."

Climatologist: "I've been to Antarctica and studied ice core data. It is all very murky near the bottom of the ice, below what we can confidently identify as the 34,000-year level. It's almost as though there had never been any ice on that particular continent…or perhaps even on the entire Earth before that time. It looks to me as though the Ice Age of 26,500 years ago I talked about last week was Earth's only Ice Age and it's getting warmer because much of that ice has already melted."

Geophysicist: "I'm concerned that the measured viscosity and speed of eddy currents in subterranean magma are not high enough to produce the force necessary to lift the peaks of the Andes to their spectacular height. Some other mechanism must have been in play."

Paleontologist: "I agree with our poor deprived 'master's degree only' friend that Argentinosaurus is too big to stand. Now what?"

Engineer: "Don't pity me, my dear PhD friends. Remember, your specialties are taught in 'The College of Arts and Science.' Engineering is taught in 'The College of Applied Science.' Therein lies a very subtle difference. 'Art' gives more leeway for interpretation and personal preferences. 'Applied' requires certain disciplines must be followed. At the master's degree level, I am not encumbered by the consensus-generating peer-review process to make sure we are all singing out of the same hymnal. I think whatever killed the dinosaurs led to the start of that Ice Age our climatologist just mentioned."

As they dispersed after lunch, the engineer said to his friends with a conspiratorial wink, "The secret to the solution to this dinosaur/global warming problem is that the Earth has changed its shape and is just now ending its one and only Ice Age! With your help and encouragement, I think I could write a book that will air all our collected concerns, as long as you four are the only peers involved in its critical review. We can do this! We'll call ourselves 'The Egg-Shaped Earth Society.'"

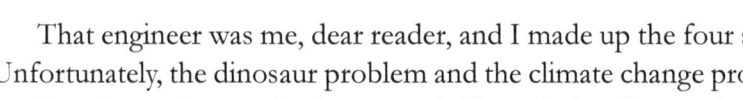

That engineer was me, dear reader, and I made up the four scientists. Unfortunately, the dinosaur problem and the climate change problem are not imaginary; they are both very real. Though the climate change issue is far more pressing, understanding the solution to the dinosaur problem is a critical first step in understanding our Solar System's evolution and appreciating the cause of the global warming problem. For that reason, let's begin with a look at those impossibly magnificent creatures.

Part 1

We've Got A Problem!

Five Friends

Chapter 1

Those Darn Dinosaurs

Why did dinosaurs have to grow to such an enormous size? If only their size had stopped at that of the Stegosaurus, the T. rex, or the Triceratops—each of which weighed about the same as a large modern-day elephant—and if only they had held their heads no higher than a modern-day giraffe, then the thinking of modern science on the subject of dinosaurs would not be challenged, and I may not have written this book. But no! They had to include in their history a group of sauropods called titanosaurs that lived about 80 million years ago, weighed 100 tons and stood 50 feet tall.

I first became aware of the dinosaur/gravity problem in an article I read on the web many years ago and was moved to do those "glazed-eyed" calculations myself to see if the largest dinosaurs could possibly stand up on Earth today (I will explain these calculations in detail in Chapter 3). The results? Unless the effective gravity on Earth during the Age of the Dinosaurs was about one third of what we feel today, they could not! This simple truth, and the problematic nature of various solutions proposed by others, led me to develop a new theory of a much simpler nascent solar system capable of providing the necessary reduced effective gravity that would allow the dinosaurs to stand, and still predict all the features of the much more complex solar system we see today.

It turns out that by proposing electricity as a much more important player in the creation of the solar system than gravity (to which the current scientific thinking gives all the credit), everything falls into place.

Who would have thought that a two-star/three-planet solar system would evolve as the solution to the dinosaur problem? Or that this unfamiliar arrangement would share a common atmosphere? Or that it would last until 34,000 years ago? Or that during that time period, humans would evolve and preserve enduring myths and legends that would provide all the clues I would need to understand what really happened?

As I write this epistle, I am still amazed at the audacity of authorship that an individual must have to even contemplate writing a book such as this. What drives a person, particularly a first time author, to write something on a subject about which they feel passionately when there are no guarantees that there will be a receptive audience for their work? Perhaps Toni Morrison had the answer when she wrote, "If there's a book you want to read…but it hasn't been written…then you must write it." To which I have to respond: There was. It hadn't. So, I did.

Fortunately for me, there are many learned authors, both in print and on the web, whose works have blazed a trail into the controversial subject matter of this book, using scores of footnoted references to other scholarly writings as a backup to their hypotheses. In their rigor, they have provided the "credibility factor" that I rely on to make believable the story I will tell. I will refer to many aspects of some of these authors' writings as I spell out a scenario that is more all-encompassing than any of their individual writings.

The result is this book, which will reveal to you a very plausible alternative to the standard model that astronomers, astrophysicists, geologists, geophysicists, climatologists, archeologists, anthropologists, and historians have been using to describe the history of our Earth, its solar system, and the universe at large. On numerous occasions during the writing process, certain "Aha!" moments occurred when I experienced the pure joy of a truth being revealed or a suspicion confirmed. I will explain these special moments as they occur in the text.

Occasionally, in the context of an explanation, I will do some simple calculations to prove a point or to determine an approximate size or distance, and the results may be presented in very specific numbers. This is not to imply that these numbers are somehow the only solution to the specific problem, rather, these numbers are simply being used to see if there is a possible solution that agrees with all the assumptions I made and that does not preclude the train of thought being pursued. There may be many other specific values that solve any of the problems I address, but for my purposes, one value that does work is sufficient.

Some of you might ask, why does an understanding of the solar system's formation matter? To which I would reply: Without an understanding of the history of each planet or other object in our solar system, there can be no understanding of our own home planet. Studying the high temperature and thick atmosphere of Venus or the lack of appreciable atmosphere on Mars to try to understand Earth's atmosphere while contemplating climate change issues is useless unless you know the solar system's history. And as future solar system exploration by robot probes or humans reveals new findings, the accuracy of their interpretation is totally dependent on the accuracy of that history.

This book is dedicated to the search for that understanding, and to all the dinosaurs that are no longer with us. Let me take you on a voyage of discovery that will lead to an understanding of a solar system with only three planets, two stars, one climate, and those darn dinosaurs who left their bones behind as an "in your face" challenge for us to resolve.

I have so much to tell you that I can barely contain my enthusiasm, but before I begin my story it is appropriate for me to provide a brief look at what I call "The Conventional Wisdom" used by mainstream scientists to tell the story of how our solar system evolved. It will provide a well-respected frame of reference for our discussions, to which proposed variations can be compared.

Five Friends

Chapter 2

The Conventional Wisdom

The Conventional Wisdom, stated here in a much abbreviated and oversimplified form, proposes that about 4.5 billion years ago, stardust collected in clouds, pulled together by mutual gravitational attraction until a central core of hydrogen atoms formed. Gravity continued to squeeze the hydrogen until it was compressed to a high enough temperature and pressure that a gravitationally confined nuclear fusion reactor was created. That reactor turned hydrogen into helium with the release of copious quantities of energy in the form of light and other fusion products. A star—namely, our Sun—was born.

But our Sun is only a medium-sized star producing only helium. For our solar system's planets to have formed, much heavier elements were necessary. In fact, all of the elements that appear today in our periodic table must have been available for the next step of the solar system's formation to occur. To provide these elements, a necessary precondition for the formation of our Sun is included in the Conventional Wisdom—namely, that a much larger star had formed and gone through several stages of evolution, first fusing helium out of hydrogen, then carbon and oxygen, then progressively heavier elements until iron and nickel were formed. Once no more fusion fuel was available, that star then collapsed in on itself and exploded in a supernova, spraying the few elements (He, C, O, Ne, Mg, Si, S, Fe, Ni) that were created therein throughout the local regions of space. The energy of this explosion fused the remaining 86 naturally occurring elements from the element debris cloud. It is from this stardust that our Sun was created.

An accretion disk formed around this solar core, made up of the supernova-created heavier elements. This debris began clumping into small rocky cores that eventually produced the four inner rocky planets and left a ring of small fragments that formed the asteroid belt. The four outer gas giant planets coalesced from lighter elements in the same accretion disk. Each planet formed in essentially the orbit it occupies today.

Some planets acquired atmospheres; others did not. On Earth, the third rock from the Sun, the moderate temperature, stable atmosphere, and the presence of water allowed life to exist. Evolution began in the oceans, expanded to the land above sea level, and proceeded in fits and starts until the collection of species we see on Earth today was formed. Each species evolved at its own slow pace. Some morphed into higher species. Most became extinct either from environmental changes or from domination by another, "more fit" species.

The planet itself evolved to its present collection of continents by the forces of internal convection currents in its molten core acting on the thin solidified crust, building mountains in certain places along the way. All of this activity proceeded at a very slow, uniform pace with only occasional violent incidents, some of which were sufficiently widespread to cause mass extinctions of up to 90% of the species alive at the time.

Conventional Wisdom has changed many times since humans first started thinking about how the world began. In the Bible, God created heaven and Earth in six days. The accepted age of the Earth has progressively increased from several thousand to several million, and now to several billion years. By the late 18th century, there were two competing schools of thought on the Earth's history. One, called *Uniformitarianism*, was essentially the same as the current Conventional Wisdom, except for the recently added nuclear fusion power for the Sun and its proposed 4.5-billion-year age. The other was a competing doctrine called *Catastrophism*, which became popular as an explanation for several dramatic geological features, such as the piles of rock and debris seen in several places all around the world. This doctrine attributed these Earth surface features to certain catastrophic events that affected the entire world.

The thinking that these changes happened in short bursts rather than over long periods of time allowed for a much younger age for the planet. The Uniformitarianism doctrine ascribed these features to worldwide glaciers slowly and methodically pushing rocks around during several Ice Ages.

Once Charles Darwin's *On the Origin of Species* was published in 1859, espousing small incremental generation-to-generation changes in a species' evolution, a much longer age for the Earth became necessary. With Albert Einstein's general theory of relativity published in 1905, a means to achieve the long time period necessary for Darwinian evolution had been found. The Sun must be nuclear-powered. Uniformitarianism won out since it made available plenty of time for surviving species to evolve into more and more complex organisms, all made possible by a central power source of nuclear fuel fusing away in the core of the Sun.

The Earth that the Conventional Wisdom produces has a uniform gravitational force that is three times greater than the effective gravity that would allow the dinosaurs to walk on its surface. Unfortunately, the dinosaurs left ample fossil evidence that they indeed were here, thus requiring that the Conventional Wisdom be considered suspect. Perhaps now is the time to discuss the dinosaur-gravity problem in more detail.

Five Friends

Chapter 3

The Dinosaur/Gravity Problem

Dinosaurs are magical. They bring out the child in all of us. I remember first learning about these long-gone mystical creatures in my earliest years of schooling and being suitably impressed, particularly with the enormous size of the treetop-browsing plant eaters and the ferocious fangs of the meat eaters. Even today, I marvel at pictures of dinosaur skeletal reconstructions on display in museums and listen again to the currently popular story of an asteroid smashing into the Earth about 65 million years ago, resulting in their demise. But even with all that scientific explanation, these animals always seemed so improbable to me that the real question I wanted answered was not how they became extinct, but how they could have existed in the first place.

They were huge! In 1987, paleontologists found the bones of a dinosaur they named *Argentinosaurus*, cataloged as one of the titanosaurs. They say that at around 100,000 kg (100 long tons), it was the heaviest of all known dinosaurs and that at 30 meters (about 100 feet) long, carrying its head about 15 meters (about 50 feet) above the ground, it was the largest land animal ever to have roamed the Earth.

Even as this book is being written, a new set of fossilized bones has been discovered in Patagonia, right next door to Argentina, that is virtually the same size as Argentinosaurus. They have named this particular beast *Dreadnoughtosaurus* because it was so big that it had "nothing to fear." Except, of course, that it and all its cousins were about to become extinct.

These and other sauropods were so big that early depictions showed them wading in ponds and marshes under the assumption that the buoyancy of water would be necessary to support their huge weight. When depicted on dry land, they were shown with their tails dragging on the ground to provide the necessary support for the tail's massive weight. Unfortunately, when footprints of these large dinosaurs were found, there was no evidence of tail-dragging. Instead, there was every indication that they did indeed walk upright on dry land.

The heaviest land animal of our time is the elephant, weighing in at about 10,000 kg (10 long tons), and the tallest is the giraffe, standing some 6 meters (about 19.7 feet) high. Argentinosaurus weighed ten times as much as an elephant and had to pump blood three times higher from its heart to its brain than a giraffe. This would be impossible today—something must have reduced the effective gravity felt on the surface of the Earth by about a factor of three during the age of the dinosaurs, so their blood vessels would not burst, their bones would not shatter, and their muscles would not tear as they tried to forage in the verdant, virtual Eden that flourished on our planet in their time. Argentinosaurus and other sauropods left their bones behind as irrefutable proof that they roamed the Earth sometime in the past, challenging us to figure out how they did it.

In today's world, animal bone and muscle have the same strength regardless of the creature's size. A mouse, a squirrel, a rhinoceros, an elephant, or a human all have the same bone and muscle strength when measured in force per unit area (i.e., pounds per square inch). A small animal like a squirrel has a much greater ratio of strength-to-weight than an elephant because, as I will show momentarily, strength increases with the square of an animal's size while its weight increases with the cube of its size.

In fact, a squirrel is about 24 times stronger for its weight than an elephant. It uses this strength to move very fast and to jump to great heights or distances for its size. A squirrel can fall 10 meters (33 feet) from a tree, land on a paved surface, pick itself up, shake off the fall, and run

The Dinosaur/Gravity Problem

right back up the tree. It is this speed, agility, and toughness that allows the squirrel to survive in its treetop environment. The elephant, however, would be foolish to even consider performing those same feats. If it fell 10 meters, it would be mortally injured. But the elephant uses its great weight and strength to pull up trees for foraging and to stomp on smaller animals that would do it harm, allowing it to survive in its chosen environment.

If it were possible to increase the strength of muscle and bone, then both the squirrel and the elephant could grow to a larger size and still maintain the agility they have at their current strength level. But alas, we carbon-based life forms are stuck with the chemical nature of our bodies. If evolution had found a way to improve upon bone and muscle strength, we surviving species would enjoy that greater strength. But we don't, and neither did the dinosaurs.

In light of these strength limitations, here is how I proceeded to analyze the Dinosaur Gravity Problem. From simple geometry, it can be shown that any animal's weight is proportional to its volume and that if an animal's length, width, and height are increased by a factor of two, its volume (and therefore its weight) will go up by a factor of eight, the cube of two.

It can also be shown that any animal's strength is proportional to the cross-sectional area of its muscle and bone and that if the cross-sectional diameter of an animal's leg is increased by a factor of two, the cross-sectional area of the corresponding muscle and bone, which is a function of their diameter squared, will go up by a factor of four, the square of two.

Consider what would happen if our present-day 10,000 kg (10 long tons) elephant were to double its dimensions. Its weight would be eight times greater at around 80,000 kg (2 x 2 x 2 x 10,000 kg = 80,000 kg) (80 long tons), close to the estimated weight of our subject dinosaur. But the available leg area to support our hypothetical doubled-dimension elephant would only be four times its original leg area, causing the stress on muscle and bone to double. Imagine an elephant spending its entire life carrying another elephant on its back. Our "2X elephant" would rend muscle and bone in any strenuous effort at that huge size, and by inference, an Argentinosaurus would similarly have been in dire straits if the gravity in its day were equal to that of our world today.

Even more troubling is the issue of blood pressure. The giraffe has the highest known blood pressure in the animal kingdom today, allowing it to hold its head as high as it does when munching on high, treetop leaves while still pumping blood to its brain without fainting. It can be shown that pressure at the bottom of any column of liquid is proportional to the height of the column. The height from the heart of an Argentinosaurus to its brain would have been about three times that of a giraffe, so its blood pressure would have had to be three times higher than that of a giraffe in order to stay alive.

Both the strength and blood pressure problems could be solved if there were a means whereby the effective gravity on Earth during the age of the dinosaurs could have been reduced to one-third of the value of gravity today. Let's call the Earth's present gravity 1G. Knowing that the gravitational force exerted at any given point in space by a body is proportional to the mass of the body and inversely proportional to the square of the distance from the center of mass of the body to that given point—and knowing that the mass of a sphere is proportional to its diameter cubed, and that the distance from a point on its surface to its center is proportional to its diameter—it can be shown that if you divide the diameter of a sphere cubed (the mass effect) by its diameter squared (the inverse square of distance effect), the surface gravity of that sphere will be directly proportional to its diameter ($D^3/D^2 = D$).

If you wanted to get one-third G on the surface of a spherical rock like the Earth, simply reduce its diameter to one-third of its present value. The present-day Earth is roughly 12,756 km (7,926 miles) in diameter, so a one-third G Earth would be about 4,250 km (2,641 miles) in diameter, with a volume only about two-thirds that of the planet Mercury. A rock that size would have only one-twenty-seventh the mass and one-ninth the surface area of the present-day Earth. There is no credible way that such a small planet could have grown to the size of the present-day Earth in the time since the age of the dinosaurs.

I am certainly not the first person to write down their thoughts on the dinosaur/gravity problem. Yet all the other authors' theories that I have examined fall far short of achieving their goal and typically offer one of four answers: reduce the force of gravity (as discussed above), increase the strength of dinosaur bones (why has that helpful evolutionary trait not been passed on to present-day animals?), use atmospheric buoyancy (pterodactyls would not have required large wings), or simply say "What problem?"

One of the references I reviewed while researching this problem was an extensively peer-reviewed study of the upper limit of the size for dinosaurs, utilizing "allometric scaling"—the phenomenon where the skeletal structure of a land animal becomes much stronger and more robust relative to the size of the body as the body size increases. For instance, the proportionately thicker bones in an elephant compared to those of a tiger are an example of allometric scaling.

The study in question is titled *"The Size of the Largest Land Animal"* by J.E.I. Hokkanen, which appeared in the *Journal of Theoretical Biology*, v.118, p.491-499, in 1986. The abstract for that article states: "The upper mass limit to terrestrial animals is studied using physical arguments and allometric laws for bone and muscle strength and animal locomotion. The limit is suggested to lie between 100,000 kg and 1,000,000 kg. A possibility for a still larger mass, in case of new adaptations, is not excluded." The lower weight limit stated is exactly the 100,000 kg that paleontologists have estimated for Argentinosaurus, the reference sauropod I have used for my dinosaur/gravity problem discussion. The upper number is ten times this weight. Any scientist questioning the huge size of this animal would be comforted by these proposed upper limits to size and simply say, "What problem?"

My basic counterargument is that when body size increases, and skeletons become more robust and massive to carry the extra load, the proportion of body mass dedicated to muscles must be correspondingly reduced. The consequence is that the larger animal is less agile. A 10,000 kg elephant is already slow and lumbering in its motions. Imagine how slowly an elephant ten times as heavy would move (or, at Hokkanen's upper limit, at 100 times as heavy) if it could even move at all.

The following chapters document my efforts to explore various means to reduce the gravity effect on large dinosaurs.

Chapter 4

Gravity and Centrifugal Force

Gravity may be invisible from our viewpoint, but there is evidence everywhere that it exists. Let's keep gravity's undeniability in mind as we imagine the creatures whose bones have been placed into evidence.

Before we dismiss the possibility that there is a solution to the dinosaur/gravity problem that uses only the forces considered in the Conventional Wisdom, there is one more scenario I would like to present to you: Can the combined effect of centrifugal force and a strong gravity gradient produce the necessary effective gravity reduction? Here is a brief overview on these two forces that all scientists agree are indeed at work on planet Earth today.

The only significant force acting between the Sun and the planets of our solar system in the present age is gravity, causing the Sun, planets, moons, asteroids, and comets to pull on each other with a force directly proportional to their individual masses and inversely proportional to their separation distances squared. On Earth, one of the more subtle of these gravitational forces can be observed in the form of tides. The Sun and the Moon pull on the liquid mass of the oceans, reshaping their surfaces due to gravitational gradients both in force and direction.

Picture the Moon directly above Earth's equator. Since the Earth is not a point mass but has a diameter of about 12,756 km (7,926 miles), and since the Moon is only 384,000 km (238,600 miles) away from the Earth, the magnitude of the Moon's gravitational effect on the Earth is

Five Friends

stronger by about 3% on the side of the Earth facing toward (and therefore closer to) the Moon, and about 3% weaker on the side facing away from (and therefore farther from) the Moon. The resulting effect is the slight reduction of the effective gravity on both the facing and non-facing sides of the Earth, causing a raising of ocean tides in both places.

In addition, the direction of the lunar gravity force at both poles is at an angle that is about 1 degree below the horizon (when the Moon is directly over the equator, you cannot see the Moon from either pole). The resulting effect is that the Moon tends to add about 1.7% of its gravitational effect on the Earth to the Earth's own gravity at the poles, tending to lower the ocean tides at both poles. The much larger mass of the Sun, at its much greater distance, has a similar and somewhat smaller effect.

These two gravity gradients of magnitude and direction constantly reshape the liquid mass of the ocean into something other than a sphere as the Sun and the Moon continually change their relative positions in the sky. When the Moon and the Sun are aligned on the same side, the observed tidal bulge is at its largest, and when they are on opposite sides, the observed bulge is at its smallest.

These same gravity gradients tend to align any non-symmetrical object in orbit around a source of gravity so that the long axis of the object points toward the source of the gravity. As the alignment of such an object slowly drifts, allowing one end of the object to move slightly closer to the source of gravity, the force of gravity will be stronger on that end, tending to pull it even closer to the source. A familiar example of this effect is Earth's Moon. It is slightly oblong in shape and has its long axis pointing toward the Earth. That's why people on Earth always see the same face of the Moon.

A second force, similar to gravity, is produced as an object rotates around another object. The centrifugal effect of rotation (tending to throw an object away from a mass around which it is orbiting) is balanced by the gravitational force pulling the two objects together. The Moon and other satellites orbiting around the Earth are responding to this balance of forces. The oceans and the entire mass of the planet also respond to the centrifugal force associated with the Earth's rotational speed about its own axis and change their shape to the extent that the Earth is not a true sphere but rather an oblate spheroid with its equatorial diameter slightly larger than its polar diameter. As a result, a man standing at the North or South Pole will weigh slightly (about 1%) more than he does when standing at sea level near the equator.

The reason for this effect is twofold. First, he is closer to the center of the Earth at the pole than at the equator, so G is greater. Second, when he is at the equator, he is farther from the center of the Earth, so G is less, and he is as far as possible from the Earth's rotational axis, so the centrifugal force of the Earth's spin is trying its best to throw him off the planet.

Having established that the present-day Earth already has a non-uniform gravitational force on its surface, my next logical step is to calculate whether combining the centrifugal phenomenon and the tidal gravity gradient forces previously discussed can create a combination of radial and axial forces strong enough to lower the Earth's effective gravity such that it does not exceed 0.33 G anywhere on the planet's surface.

To achieve this desired effect, I set up the problem as follows: First, position the Earth such that its spin axis is pointing at the Sun (the strongest source of gravity available in the solar system), and determine the spin rate that would cause centrifugal force to reduce the effective gravity at the equator to no greater than 0.33 G. Then, determine the distance with respect to the center of the Sun's mass where the resulting gravity gradient produces no greater than 0.33 G at the poles.

Since a satellite in low Earth orbit has a period (the time it takes to circle the Earth) of 1.5 hours, it follows that if I spun the Earth at a rotational speed of one revolution every 1.5 hours, I would create 1 G of centrifugal force at its surface to completely balance the 1 G of gravitational force. Since I only need to overcome 0.66 G to get a net 0.33 G force at the equator, and since centrifugal force is inversely proportional to the square of the period, the Earth would need to spin at one revolution every 1.85 hours—or about 13 times the Earth's current rotational speed —to get the desired effect.

To maintain the Earth's spherical shape, the pull on the poles from the Sun's gravity gradient would have to be strong enough to generate 0.66 G of lift. To get a gravity gradient this strong, my calculations show that the Earth would have to be about 660,000 km (410,110 miles) from the center of gravity of the Sun. Unfortunately, the diameter of the Sun is 1,390,000 km (863,700 miles), making its radius 695,000 km (431,850 miles). This means that the Earth would have to be orbiting inside the Sun, a physical impossibility. In addition, since a spherical Earth configuration has no gravity gradient stabilization, it would quickly become unstable.

This exercise has obviously not solved the 0.33 G surface gravity for a spherical Earth, but it has shown that there is a means to apply separate and adjustable radial and axial forces to the Earth that can affect its effective surface gravity. If I increase the solar gravity gradient effect, the Earth could be distorted into an egg-shaped prolate spheroid with a sufficiently elongated shape to allow the gravity gradient to keep Earth in a stable orientation with its spin axis pointed at the Sun. Because such a shape is non-symmetrical, the effective surface gravity will not be uniform over its entire surface. However, if I adjust the two forces so that 0.33 G is obtained at the equator, the G at the pole will be lower due to its greater distance from the center of the Earth, and the result will be a planet where, though not uniform, gravity does not exceed 0.33 G anywhere on its surface.

Unfortunately, both of these reduced-gravity scenarios place our planet much too close to the Sun. Fortunately, what I have been able to show is that as long as I can pull on a planet with separate axial and radial forces, I can obtain a stable prolate spheroidal shape and the reduced G level I have been looking for. But alas, in our solar system, gravity and centrifugal force are not a means to that end.

Gravity and Centrifugal Force

I had one of the many "Aha" moments (let's call this one "Aha! #1") revealed to me at this point in my writing when I realized: A spherical Earth and dinosaurs are incompatible.

Aha! #1:
A spherical Earth and dinosaurs are incompatible.

That realization reinforced an earlier observation that I and several other authors whose work I have reviewed had made—namely, Earth's primordial continent named "Pangaea" doesn't fit on the spherical Earth of today, but does fit on one end of a prolate spheroid of the same volume. Perhaps it is time to consider whether the shape of Pangaea is a clue to the solution to this problem.

Five Friends

Chapter 5

Pangaea

Ever since the first maps of the world were drawn centuries ago, it became apparent that the eastern and western shores of the Atlantic Ocean could be fitted together like two pieces of a jigsaw puzzle. It appeared as though the continents on opposite shores were once touching, and had somehow drifted apart. It was as though these giant slabs of rock were trying to tell us they once formed a single continuous landmass. The idea was that on this supercontinent, today's Atlantic Canada was connected to Spain and Morocco.

In the early 1970s, after extensive mapping of the floors of all the world's oceans, which included the discovery of a Mid-Atlantic Ridge, the theory of continental drift gained renewed credibility in the form of a new science called Plate Tectonics. With this new mapping information, all the continents of the Earth could be joined together at their continental shelf boundaries to form one universal landmass called "Pangaea."

But in order to fit Pangaea on the present spherical Earth, that essentially hemispherical primordial landmass had to have one large piece missing, as though someone had taken one slice of pie from the whole. That gap was named the "Tethys Sea."

The Conventional Wisdom has no problem with this arrangement since it contends that Pangaea was preceded by a series of older continents that had been bumping into each other since the Earth's crust first cooled—they just happened to have all collided at one time about 300 million years ago to form Pangaea.

If that pie-shaped gap were closed to form the true unbroken hemisphere that one might expect for the first newly formed landmass on a freshly minted planet, it would no longer fit on our 12,756-km (7,926 miles) diameter sphere, but it would fit very nicely on a sphere that was closer to 11,000-km (6,835 miles) in diameter.

Unfortunately, this alternate Earth diameter does not solve the 1/3rd G problem. An 11,000-km (6,835 miles) diameter sphere is 86% of the present Earth diameter, and following the logic I discussed earlier, it would have only 86% of Earth's present gravity. So, yes, it helps, but nowhere near enough to allow the largest dinosaurs to exist.

In addition, to bring such a planet up to the volume of the present Earth would require adding the equivalent of 2.5 Mars volumes or a volume equal to 200 of Earth's Moon. Any celestial event that attempted to add that much volume to a primordial Earth would certainly have caused it to shatter, destroying the planet we now live on. So it turns out that the 11,000-km (6,835 miles) spherical Earth does allow an uninterrupted Pangaea, but it is not a viable configuration to solve the basic large dinosaur problem.

I asked myself, *What if I tried to place Pangaea on a different shape Earth?* I started to picture it placed on the prolate spheroid I described earlier. Such an Earth could have curvature at the poles similar to the 11,000-km (6,835 miles) sphere that allows Pangaea to fit so nicely, and yet have enough volume to contain all the mass of the present Earth. I ran some numbers that show an ellipsoid with an equatorial diameter of 11,054 km (6,867 miles) and an 18,182-km (11,298 miles) pole-to-pole dimension would satisfy these two requirements. If the Earth had this shape during the age of the dinosaurs, it could slowly reshape itself into a sphere between then and now without shattering. Granted, there would be major surface cracking and wrinkling since the surface area of an ellipsoid is greater than the surface area of an equivalent volume sphere, and all that extra area has to go somewhere—but it could be done!

This is a far more encouraging option than the 11,000-km (6,835 miles) diameter sphere previously considered since the ellipsoidal-shaped Earth would allow an uninterrupted, approximately hemispherical primordial landmass and mass to be positioned at one of the poles, This Earth could eventually change its shape to a sphere without changing its volume.

Unfortunately, it still does not solve the 1/3rd G problem, since the gravity at its poles would be 0.50 G, and the gravity at the equator would be 1.33 G. In addition, it introduces a major new problem, namely that the ellipsoidal shape could not have formed in the first place unless there were additional forces acting on the Earth from the time it solidified through the time of the age of the dinosaurs. To resolve these issues, it is our task to determine the source of those additional forces.

Five Friends

Chapter 6

Electricity

How can we make sense of the Dinosaur Age?

Electricity is the only other gravity-like force currently at work in the universe—similar in that they are both forces that act between objects without direct contact. The case I will present below proposes that electricity is the source of the distortion force necessary to get the desired surface gravity and the desirable Pangaea-friendly planet shape that (when combined) allows for the Dinosaur Age to make sense. Information from the Sun and the solar system, collected on Earth using powerful land-based telescopes, orbiting satellites, and space probes, shows a solar system that is extremely active electrically.

We see huge solar flares arching up to stupendous heights above the Sun's surface—heights that dwarf the size of our entire planet. We see twisting electrical currents in the Sun's corona and at the edges of sunspots in the Sun's photosphere, which provides our planet with its life-giving sunlight. We see comet tails that point away from the Sun both on their approach to and on their departure from close encounters with the Sun. With awe, we view the aurora borealis in the northern nighttime sky caused by electrically charged cosmic particles blasted from the Sun as they bombard the magnetosphere of our planet. We see an electromagnetic flux tube that extends from one of Jupiter's poles to the other while enveloping Jupiter's closest moon, Io. There is no doubt that we live in an electrically active solar system, so the possible role of these electrical forces in shaping the Earth and its effective gravity is one of the clues that we need to explore.

Compared to electrical forces, gravity is a wimp. It is impressive to note that two protons, each with a single positive electrical charge, repel each other with a force that is 10^36 (that's a 1 with 36 zeros after it: 1,000,000,000,000,000,000,000,000,000,000,000,000) times as strong as their mutual gravitational attraction trying in vain to pull them together. One way to see these electrical forces at work is to perform the simple child's experiment of building up a static electrical charge on a balloon by rubbing it against a cloth, touching it to the ceiling, and watching it stay there, defying the entire mass of the Earth trying to pull it down. It's a trick that never ceases to amaze even the adults in the room and provides inspiration for the next step in our riddle-solving journey.

As part of our consideration of electricity as the solution to our planet shape problem, it is important to review the role of plasma in the universe. Plasma is the most abundant form of matter in the cosmos. It is the fourth state of matter, along with solid, liquid, and gas, which comprise the other three. It is composed of positively and negatively charged particles. The positive charges consist of ions in the form of atoms and molecules that have a deficiency of electrons. The negative charges consist of either free electrons or negatively charged particles, such as ionized atoms and molecules, that have extra electrons. Like gas, plasma can be enclosed in a container. Plasma in this form can be seen every day in such things as neon signs and fluorescent light bulbs. But unlike gas, electrical currents and the magnetic fields they create can also shape plasma, as seen in solar flares, corona filaments, and flux tubes.

Plasma has been studied in laboratory experiments using a clear-glass tube with anode and cathode electrodes positioned inside at opposite ends. When the tube is filled with plasma at a very low pressure, similar to what would be found in the vicinity of the Sun and its solar system, and when an applied voltage between the electrodes is increased above a critical level, current will begin to flow. But unlike current flowing in a wire where there is a uniform voltage drop between one end of the

Electricity

conductor and the other, the voltage drop in the contained plasma occurs in discrete steps at what are called Double Layer (DL) points that form at regular intervals along the tube's length, with virtually no voltage drop occurring from one DL point to the next.

Individually, DL points have a finite thickness and are positively charged to one voltage on the side facing the anode, and negatively charged to another voltage on the side facing the cathode. If there were four DL points between the anode and the cathode, each DL point would have 25% of the total voltage drop imposed over its very slight thickness. The resulting local voltage gradient within each DL point is extremely high.

With this rudimentary understanding of plasma in mind, I asked myself, *What if our Sun had a much smaller companion star early in its cosmic history?* This would not be unusual since at least half of the observable stars in our galaxy are part of a double-star system. Some are part of three-, four-, or even five-star systems. If the Sun and this hypothetical companion star were electrically charged to a sufficiently large difference in potential, a strong electrical field would exist between them.

Electrical currents would begin to flow in the conductive plasma that permeates the space between the two stars, producing magnetic fields that would tend to shape the currents into a twisting helix known as Birkeland currents, most often occurring in pairs. These currents tend to compress any material, ionized or otherwise, into the confined space contained within the helices by a process known as the "Z-pinch" effect. The outer surface of this resulting flux tube would carry current with the same continuous voltage drop between the Sun and its companion star as would be experienced in a copper wire strung between them.

Current also flows through the volume of plasma trapped within the flux tube, but in a manner similar to the lab experiment tube I discussed earlier, with discrete DL points developing at intervals along the length of the tube, producing high voltage gradients at each point and virtually no voltage drop in the space between points. Material within the flux tube that was compressed by the Z-pinch effect, tends to collect within the DL points.

Although a flux tube between stars would be unimaginably larger than the simple laboratory plasma experiment just discussed, DL points of planetary proportions, capable of enclosing a planet within their thickness, would still form. Fortunately for my quest, the electrical voltage gradient inside a DL point acts essentially the same as a powerful gravity gradient, allowing it to produce the axial forces necessary to stretch a planet-sized body to an ellipsoidal shape. Additionally, the electrical charge in the surrounding Birkeland current flux tube can create radial electrical forces acting in the same direction that centrifugal forces act on a rotating body. The result is that, when properly balanced with the DL axial forces, the reduced surface gravity effect I am looking for can be provided.

Figure 1: "Flux Tube Voltages" shown below is my schematic representation of what such a flux tube might look like. The circle on the left represents the Sun, drawn to the same scale as the proposed companion star represented by the small circle on the right. However, the relative distance between them is reduced and not shown to scale. A flux tube with four DL points, surrounded by a single spiral Birkeland current, is shown connecting the two stars.

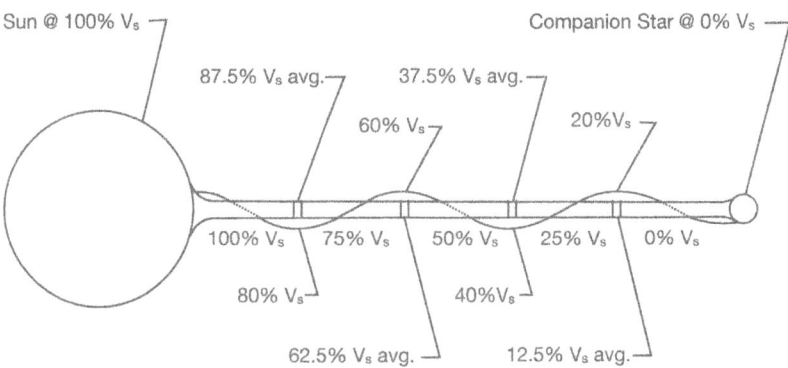

Figure 1: **Flux Tube Voltages**

Electricity

Suppose that the Sun is charged to a positive voltage (Vs) with respect to the companion star. Since voltage drop within the flux tube can occur only at the four DL points, 25% of the total voltage difference will occur across each point. As depicted in the sketch, the DL point closest to the Sun has its left side at 100%Vs and its right side (facing the companion star) at 75%Vs. The next point has 75%Vs on its left side and 50%Vs on its right side. The next point has a corresponding 50%Vs to 25%Vs split, and the point nearest the companion star has a corresponding 25%Vs to 0%Vs split.

The spiraling line in the sketch represents a typical Birkeland current, one of a pair or more that form the cylindrical boundary of the flux tube. Remember that each Birkeland current acts such that the voltage drop for its portion of the current flow between the companion star and the Sun will be evenly distributed along its length. At the Sun, the voltage is 100%Vs. Moving toward the companion star, the voltage drops to 80%Vs at the DL point closest to the Sun. Then, as each successive DL point is passed, the voltage drops to 60%Vs, then 40%Vs, then 20%Vs, and finally to 0%Vs when the companion star is reached.

If a planet-sized object were placed within the DL point closest to the companion star shown in Figure 1, it might look somewhat like the arrangement in Figure 2: "A Planet in a Flux Tube DL Point."

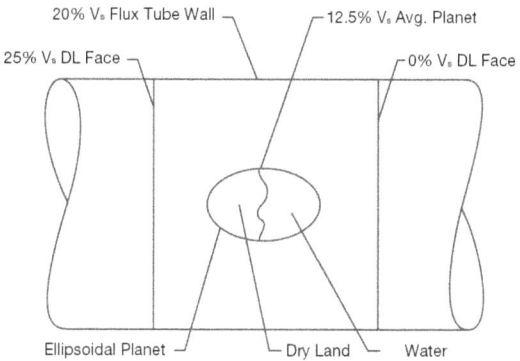

Figure 2: **A Planet in a Flux Tube DL Point**

The horizontal lines represent the flux tube's outer diameter with Birkeland currents flowing. The vertical lines represent the two faces of the DL point. The oval in the middle represents a planet -sized object. From the discussion of the previous sketch, I have shown that the left side of the DL point is at 25%Vs and the right side is at 0%Vs. The voltage on the planet-sized object will be at an average of the two DL point voltages, or about 12.5%Vs. The difference in voltage between the two DL point faces and the planet will tend to stretch the planet to an egg shape, as shown in the figure. The voltage of the flux tube wall will be about 20%Vs and will exert a radial pull on the enclosed planet at its voltage of 12.5%Vs.

One of the "Aha!" moments I had during my research and writing occurred when it became clear that the voltage in the Birkeland currents flowing in the wall of the flux tube would be different from the average voltage on a planet trapped in the local DL point, allowing a radial electrostatic force field to exist—a necessary condition for the 0.33 G maximum surface gravity I was looking for. Thus, Aha! #2: Different voltages in the wall of the flux tube and the DL point cause radial and axial forces capable of creating an egg-shaped planet.

Aha! #2:
Different voltages in the wall of the flux tube and the DL point cause radial and axial forces capable of creating an egg-shaped planet.

Voila! I have done what I set out to do. I have found a credible means to pull on a planet with separate axial and radial forces, other than gravity and centrifugal force, with sufficient force to provide a symmetrical primordial continent shape and the surface G level necessary for the existence of the largest dinosaurs.

For this revelation to be credible, there must be an explanation for how our Sun came to possess a companion star. The Conventional Wisdom briefly described earlier has no such explanation. It must be abandoned and replaced with a new unconventional wisdom that turns out to provide a remarkably simple solution.

Chapter 7

The New Unconventional Wisdom

Alas, we can't go back in time for a glimpse of the birth of our world, so making the best guess about what happened has become an obsession for mankind through the centuries. For the hypothesis about this which I have been developing to be proven true, a new model that replaces the Conventional Wisdom is required—one that does not rely solely on gravity for the formation of the solar system. Unfortunately, anyone with a good education has been taught the current Conventional Wisdom by credentialed authorities, and since none of us were around to witness what actually happened, we take them at their word. But there are those who question things that don't seem to fit. They are the ones who start developing alternate approaches to problems until a solution arises that fits better than the Conventional Wisdom. I consider myself to be one of them.

I like to refer to this new approach as the "Unconventional Wisdom." It repudiates so much of what we all have been so carefully taught that many reading this dissertation will find it too much to accept. Albert Einstein famously said, "If at first an idea is not absurd, there's no hope for it." If indeed this axiom's corollary is true, then I'm really on to something. Fortunately for me, there is a core of individuals who have already discovered the essence of this alternate explanation of the universe, finding it so much more satisfying and problem-solving than the Conventional Wisdom that they embrace it wholeheartedly. We hope you'll join us.

The most revealing recent dissertation on the subject at hand is an extensively referenced and footnoted book by Donald E. Scott, titled, *The Electric Sky*, with the subtitle, *A Challenge to the Myths of Modern Astronomy*, published in 2006. With his background as a professor of electrical engineering at the University of Massachusetts/Amherst, Dr. Scott is a well-qualified standard-bearer for the Electric Universe/Plasma Cosmology that is emerging as a credible alternative to Conventional Wisdom and its heavy reliance on gravity to explain the cosmos.

His book challenges Conventional Wisdom by providing electrical explanations for every observed cosmic phenomenon. He explains that the redshift calculations determining the receding speed of distant galaxies miss an additional redshift caused by electrical currents in those same galaxies, thereby rendering the calculations of the age of the universe in error by a significant amount on the high side. With his guidance, readers come to realize that there is no need for black holes, neutron stars, missing mass, dark matter, WIMPS, the Big Bang, or the menagerie of strange phenomena that the Conventional Wisdom relies on to support its version of how the cosmos works.

To me, his most convincing argument for an electrically powered Sun, as opposed to the Conventional Wisdom's nuclear-powered Sun with its requisite six-million-degree Kelvin core, is the readily observed phenomenon of sunspots that appear as holes in the photosphere. These openings reveal a dark 3,800 K to 4,000 K surface below the 5,800 K photosphere surface. If the Sun were being heated from within, the temperature below the photosphere would be *hotter*, not *cooler* than its surface. I am amazed that this simple observation was not sufficient to derail the "nuclear-powered Sun" theory in its infancy.

In *The Electric Sky*, Dr. Scott lists the pioneers in the field of Plasma Cosmology, including: Benjamin Franklin (1706–1790), with his discovery that the sky was essentially electrical in nature; James Clerk Maxwell (1831–1879), with his equations for all things electrical and magnetic, including the electromagnetic spectrum; Kristian Birkeland (1867–1917), with his field studies of auroras (northern lights) in Norway and their connection to

The New Unconventional Wisdom

electrical currents from the Sun; Irving Langmuir (1881–1957), for his work in electrical plasma; Hannes Alfvén (1908–1995), for his understanding of the large-scale filamentary structure of the universe; Anthony L. Peratt for his understanding of the plasma universe and the formation of systems of galaxies; and Wallace Thornhill (1942-2023) for his understanding of the importance of electromagnetism in celestial dynamics and his attitude that "…the lessons learned best are those where we think for ourselves."

Dr. Scott, Wal Thornhill, and many other like-minded scientists are also involved in The Thunderbolts Project, which presents videos on The Thunderbolts Project YouTube Channel, featuring lectures and astronomical observations of electrical phenomena in our universe, and in the SAFIRE Project, which performs controlled laboratory experiments to reproduce and closely examine many aspects of the Electric Universe.

I am indebted to all these individuals as I present the lessons I learned while thinking for myself. Those lessons form the heart of this book. The electrical behavior of the Milky Way galaxy, the Sun, and our solar system that Dr. Scott describes are sufficient for the purpose of laying out a credible explanation for how our Sun acquired the companion star necessary for my basic hypothesis of a low-gravity ellipsoidal early Earth. This book relies on the reasoned description of a new Unconventional Wisdom that has its foundation in the works of Dr. Scott and several other authors, who will be identified as their individual contributions to this effort are brought into the story. The remainder of this volume defines my version of the new Unconventional Wisdom, starting with how the solar system really formed. Buckle your seatbelt—you're in for a wild ride. This is Catastrophism at its finest.

Five Friends

Part 2

The Rise & Fall Of The Dinosaurs

Five Friends

Chapter 8

A Solar System Is Born

Electricity is constant and relentless...thank goodness. I will begin this discussion of the new Unconventional Wisdom with a description of how the solar system was formed, starting with the fundamental understanding that the cosmos is not empty. It is filled with plasma, punctuated by galaxies and remnants of galaxies, stars and remnants of stars, planets and remnants of planets, all of which are the consequences of a relentless flow of electricity. That flow is observably filamentary (stringy) at every size level, from planetary atmospheres, through the Sun's corona, through groups of stars, even galaxies and galaxy clusters, all caused by electrical currents and the magnetic fields they produce.

Between every visible object in the cosmos, there is unseen plasma made up of ions and electrons, and occasionally a few uncharged atoms, all milling about with no particular order. But since ions attract or repel each other depending on their charge, it doesn't take long before they start migrating toward opposite charges and away from like charges. When charged particles move, they are the essence of an electrical current, and when electrical currents flow in plasma, they create magnetic fields that shape the very plasma they are made of, as is seen in the curved spirals of the Birkeland currents I have discussed previously.

Depending on the current density, which is the amount of current flowing through a given area of space, the plasma will either be not visible (called the dark current mode; i.e., radio waves), will be seen to glow (called the normal glow mode; i.e., a neon sign or a comet's tail), or will radiate brilliantly (called the arc mode; i.e., lightning, an arc welder's torch, or the visible surface of the Sun).

A long time ago, perhaps 3 billion years ago, electrical current was flowing in a cloud of ionized gas in one of the spiral arms of a galaxy that would become known as the Milky Way by a species known as humans, who would occupy a soon-to-be-created small planet they would call Earth, orbiting an about-to-be-created star they would call the Sun.

As the current flowed, spiral Birkeland currents developed, creating a Z-pinch effect that tended to squeeze dust, hydrogen, and ionized gas between them. The result was an incredibly large galactic flux tube filled with the trapped material. Double Layers (DLs) formed at increments measuring in light years along the length of the tube. Most of the voltage drop in the otherwise extremely conductive plasma occurred at these DL points, which subsequently tended to act as collection points for the material in the tube. Gradually, the material clustered together through mutual gravitational attraction, creating a sphere with a diameter of 1,390,000 km (863,700 miles).

This huge and massive ball of matter did not squeeze its contents to the temperature and pressure required for nuclear fusion. Instead, its extraordinary gravitational pull offset the nucleus of each of its atoms ever-so-slightly toward the center of the sphere with respect to the individual atom's electron cloud, thereby creating an outward-facing negative bias at each atom. The resulting electrostatic field swept any weakly bonded electrons toward the outer surface of the sphere, leaving only positively charged ions behind. The repelling forces between the like-charged positive ions exactly balanced the gravitational attracting forces between ions, resulting in a sphere with a stable core of relatively uniform density.

Just as the point at the end of a lightning rod tends to concentrate the electrical field of an impending lightning strike, this sphere of positively charged ions—tiny by galactic standards—tended to focus the cosmic electrical current flowing within the galactic flux tube's local DL onto

A Solar System Is Born

the surface of the sphere. The equatorial plane of the sphere acted like the rotor disk of a homopolar motor, allowing the flux tube's current to apply a torque at the sphere's equator, causing it to spin, dragging the remainder of the sphere along at a progressively slower rotational speed near the poles. The current density was high enough to bring the plasma surrounding the sphere to the arc mode, heating it to 5,800 K and creating an intensely bright light. A star, eventually named the Sun, was born.

Within the tightly wound current filaments in the arc-mode plasma at the outer surface of the Sun's photosphere, local Birkeland currents created extremely strong z-pinch effects that were capable of squeezing hydrogen to fusion temperatures and pressures, thereby creating helium and other fusion products and releasing even more energy into the radiation from the Sun. For hundreds of millions of years, the Sun shone brightly in the galaxy, and over time an atmosphere consisting of every molecular weight element now known was formed in and around these localized magnetically confined fusion zones that were spread over the Sun's entire surface. This same fusion technique, with the help of superconducting electromagnets, was at the heart of an early effort to create controlled nuclear fusion here on Earth in the 1950s.

During the time that it took the Sun to create its atmosphere of elements, the Milky Way galaxy itself turned about two revolutions on its own axis. Stars continually drifted in and out of the galaxy's spiral arms, between more highly charged areas and less highly charged areas. Perhaps 2.5 billion years ago, the Sun drifted into an area that was more highly charged than the area in which it had been formed. As a result, the current in the galactic flux tube containing the Sun increased to a level where the arc-mode current flow at the Sun's photosphere became unstable, causing a large area of the Sun's surface to blow off in a solar fission event, carrying with it the majority of the electrical charge from the Sun's surface, but only a small portion of the Sun's total volume. The bulk of the resulting debris coalesced into a small companion star about 165,000 km (103,000 miles) in diameter and only about 1/500th of the Sun's volume. The Sun's diameter, at 1,390,000 km (863,700 miles), was only slightly smaller than it had been before the explosion, and even though its surface area was reduced, the addition of the companion star provided enough new surface area that, with two points on which to focus, the local current density decreased, allowing a return to the stable arc mode on the surface of the Sun and to the normal glow mode on the new companion star.

The dramatic voltage difference between the two stars of the newly formed binary pair resulted in an interconnecting electrical field that acted to settle the new companion star's orbit around the Sun so that its rotational period was equal to the Sun's rotational period of 25.3 days. That same electrical field also caused Birkeland currents to flow and magnetic fields to develop, thereby creating a new flux tube with multiple DL points between the Sun and its new companion star. A somewhat similar, though significantly smaller, flux tube exists today between Jupiter's north and south poles, enveloping its closest moon, Io.

Most of the elemental particles ejected from the Sun's fusion zone were trapped within the confines of the newly formed flux tube, as if they were in a laboratory flask. The portion of the Sun's original mass that did not coalesce into the companion star or become trapped in the flux tube was blown out into space, forming the Oort Cloud, consisting of millions of icy rocks and comet nuclei positioned more than a thousand times farther from the Sun than Pluto's present-day orbit.

The wrap-around spiral geometry of the Birkeland currents acted as tension elements to resist the flux tube pressure, allowing a reasonably high density for the gases trapped therein. The voltage drop between adjacent electrons in the Birkeland currents provided the slight electrostatic force necessary for that tension to exist. The material in the flux tube was heated both by mechanical compression and by electrical current flow to temperatures that allowed many chemical compounds to form, creating the solar system's earliest common atmosphere. After a sufficient cool-down period, these hot elemental gases began to coalesce. The resulting condensate collected primarily between the layers of three of the DL points along the tube that were positioned far enough away from the Sun, and not too close to the companion star.

A Solar System Is Born

The space within these three just-right points was subjected to two strong voltage gradients: an axial voltage gradient formed by the difference in voltage between the faces of the DL that trapped the hot condensing elements, and a radial voltage gradient created by the confining walls of the surrounding flux tube (see Chapter 6, Figure 2). Though gravitational forces were trying to form the collected molten matter into a spherical shape, the combined electrostatic forces and a slight gravity gradient field from the Sun and its new companion star combined to distort the gravitational spheres into three ellipsoidal shapes that were in the process of becoming three rocky planets.

Each planet's gravity did succeed in segregating the collected material by density, causing the heavier compounds to sink, forming a molten, mostly metallic core, while the lighter mineral compounds floated on the new core's surface.

Each of these evolving planets acted as the rotor disk of a homopolar motor, in much the same way as the Sun's equatorial plane had when it was formed. As a result, the newly formed flux tube's electrical current applied a torque at each planet's equator, causing it to spin with a rotational period of 24 to 25 hours, with its rotational axes aligned closely to the axis of the flux tube. The result was a just-formed early solar system consisting of a Sun spinning on its axis, a companion star spinning on an axis parallel to the Sun's axis, and an interconnecting flux tube with three ellipsoidal planets inside, all sharing a common atmosphere, and each spinning on an axis with one pole pointing at the Sun and the other at the companion star.

Five Friends

Chapter 9

The Dimensions of the Early Solar System

Who's starring in this show that premiered in Chapter 8, "A Solar System is Born"? I'd like to pause for a moment and take stock of the newly formed solar system I described, and while doing so, I'll give some names to the character actors I have created. In the center of the drama is the principal actor, the massive star…the Sun. The co-star, if you will, is the companion star at the other end of the ethereal flux tube. For reasons that will become clear later in our story, I'll call this co-star "Proto-Saturn." The size of Proto-Saturn has been determined by combining the volume of the solar system's current planets: Jupiter, Saturn, Uranus, and Neptune. The reason for choosing this size for Proto-Saturn is that, eventually, a catastrophic event will destroy the flux tube and cause Proto-Saturn to break up into these four giant planets. More about this event anon.

In addition to the two main actors, there are three-bit players that are, by comparison, almost inconsequential in size, but vitally important to this story. These players are the three ellipsoidal planets I have just described that have taken up residence between the faces of strategically located DL points along the length of the ubiquitous flux tube, sharing its atmosphere. They are, in order of their position, starting with the planet closest to the Sun: Earth, Mars, and what I call "Proto-Venus," again for reasons that will be explained later in this book. Both Earth and Mars had essentially the same volume as they have today. The size of Proto-Venus has been determined by combining the present-day volumes of Mercury,

Venus, Pluto, all the asteroids, all the moons of all the planets that can be seen today in the solar system, and another near-Mars-sized planet that I will call Phaeton, which is no longer a part of the solar system. This makeup of Proto-Venus is another strong hint as to how the plot will develop. Stay tuned.

So far, I have developed a scenario that uses electrical forces to reduce the effective gravity on Earth's surface so that huge dinosaurs could stand erect—and to distort the Earth so that Pangaea could fit without a gap. Back in Part 1, Chapter 4, "Gravity & Centrifugal Force," I discussed how these two forces could create the planet shape and surface gravity I was hypothesizing, but only at a distance that was much too close to the Sun. In Part 2, Chapter 8, "A Solar System Is Born," I demonstrated how, by adding a companion star to the mix and connecting it to the Sun with a flux tube, I can get strong axial and radial electrical forces. These could create the same effective gravity-reducing and shape-distorting effect that the high gravity gradient found at the Sun's surface, when combined with the centrifugal force of a high rotational speed, was capable of producing but doing so at a safe distance from the Sun.

The axial component of these electrical forces would have extended like a tension cable from the Sun, through the Earth, Mars, and Proto-Venus, all the way to Proto-Saturn, holding the Sun's companion star in a tighter-than-usual orbit by supplementing the Sun's gravitational force.

Picture an Olympic hammer throw event with the athlete as the Sun, the hammer ball as Proto-Saturn, and the cable connecting the ball to the hand grip as the axial force produced within the flux tube. The magnitude of that axial electrical force must meet two requirements. First, it must provide a lifting force strong enough to reduce the effective gravity at the early Earth's poles to something less than 0.33 G so that the dinosaurs could stand up on Earth's surface. Second, when combined with the Sun's gravitational force, it must be strong enough to hold Proto-Saturn in its Early Solar System orbit.

To calculate these flux tube forces, I must first determine the most likely orbital radius for Proto-Saturn as it circled the Sun. The calculations to determine this location, and then to verify that there was sufficient force available to keep Proto-Saturn orbiting at that radius, are relatively straightforward but a bit tedious and may be skipped if so desired. The next eight indented paragraphs are dedicated to that calculation. For simplicity, I use only metric units for these calculations. The bottom line is that Proto-Saturn's initial orbital radius was 31.8 million km (1.97 million miles), and there was sufficient force to hold it there. The numbers worked, making this calculation and its results one of my most rewarding "Aha!" moments in the writing of this book.

Aha! #3:
Electrostatic forces necessary to hold Proto-Saturn in orbit and to stretch Earth are compatible.

The Calculations

Orbital mechanics provides a simple equation that will determine the radius to an object that is in a circular orbit around the Sun with a given period of rotation when gravity is the only attractive force under consideration. In this book, I refer to that radius as the "gravity-only radius." Orbital mechanics also provides another simple equation that can be used to determine the velocity at any given distance from the Sun that will allow an object to escape the gravitational pull of the Sun so that it will be ejected from the solar system. This velocity is commonly called the escape velocity.

Suppose Proto-Saturn had been orbiting at a distance from the Sun that was greater than the gravity-only orbit calculations would allow at my proposed period of rotation. And suppose also that the resulting velocity imparted by that larger radius was equal to the escape velocity from that proposed orbit. Under these conditions, if some event caused the flux tube to be destroyed, resulting in the disappearance of the axial electrical forces, and if the same event caused Proto-Saturn to fracture, creating Jupiter, Saturn, Uranus, and Neptune in the process, these four newly minted planets would all be moving at close to the escape velocity for Proto-Saturn's original orbit.

If there were also some electrical forces involved that acted to gradually scrub off just enough velocity from each planet that they did not actually escape but instead settled into their current orbits far away from the Sun, then I have a scenario that will allow me to calculate the necessary preconditions for the location of Proto-Saturn before the flux tube disappeared.

Earlier, I deduced that Proto-Saturn was locked into a rotational period around the Sun that is equal to the Sun's current equatorial rotational rate of 25.3 days per revolution. If I simply determine the radius where this particular rotational period produces a velocity that is equal to solar system escape velocity, I will have the primordial orbital location of Proto-Saturn.

The escape velocity from any circular orbit around the Sun is equal to a known constant multiplied by the reciprocal of the square root of the orbital radius. I also know that the velocity of an object in a circular orbit is equal to the circumference of its orbit, which is two times Pi times the orbital radius, divided by the orbital period, which is the time for one complete revolution around the Sun. Setting these two velocities equal, and using the present-day equatorial rotational rate of the Sun, I can solve for the orbital radius. Those calculations result in a radius of 31.8 million km and a velocity of 91.4 km/sec. This radius puts Proto-Saturn outside of the 25.2 million km gravity-only orbit that a planet with a 25.3 days per revolution period would have, but (as a frame of reference) well within the 57.9 million km present-day orbit of Mercury.

The next order of business is to determine the magnitude of that portion of the flux tube's axial electrostatic force that actually intersected the cross-sectional area of the Earth on the Sun-facing side. The value I am looking for must produce a lifting effect equivalent to 0.66 G, thereby reducing the effective G force on Earth's surface to approximately 0.33 G. I calculated that the portion of the force would have to be approximately

1.93 x 10^25 N (N is the symbol for Newtons, the metric unit for force), with a corresponding force applied in the opposite direction to the side facing away from the Sun as well.

Since Proto-Saturn's 31.8 million km orbital radius was well outside the 25.2 million km theoretical gravity-only 25.3-day orbital-period radius, the total magnitude of the electrical forces within the flux tube must have been sufficiently strong so that the combined solar gravitational force and flux tube axial force could balance the centrifugal force of Proto-Saturn in that orbit. I calculated that the Sun's gravity applied a force of 3.50 x 10^26 N on Proto-Saturn in its early orbit, and that the total centrifugal force acting on Proto-Saturn was 6.97 x 10^26 N. The difference between these two forces, namely 3.47 x 10^26 N, must have been supplied by the total flux tube axial electrostatic force. This force is 18 times greater than the 1.93 x 10^25 N portion of the flux tube axial electrostatic force needed to reduce Earth's effective gravity to 0.33 G.

It's a good thing that the total flux tube force was greater than the force necessary to perform its magic on the Earth's effective gravity, or my entire thesis would go up in smoke. The ratio of 18:1 for the available-force to G-effect-force ratio provides a comfortable margin to accommodate any error in my early solar system dimensional estimates. If, as I suspect, the axial electrostatic force distribution across the full diameter of the flux tube was relatively uniform, this 18:1 ratio can also be used to estimate the diameter of the flux tube itself.

It can be reasoned from this ratio that the cross-sectional area of the Earth on which the flux tube forces act is 18 times smaller than the cross-sectional area of the flux tube. Since the diameter of a circle is proportional to the square root of the circle's area, the diameter of the flux tube would be 4.82 (the square root of 18) times the Earth's 12,800 km diameter, or 54,000 km.

Whew! That was a lot of numbers to throw at you, but I had to do this calculation to see if my thesis was correct.

Back to the Show

Figure 3, shown below, is a sketch of the newly defined early solar system. The circle on the left represents the Sun, and the smaller circle on the right represents Proto-Saturn.

Earth, Mars, and Proto-Venus are shown as very small ellipses within a flux tube connecting the two stars. The location of the three planets within the flux tube is not arbitrary. Their relative distances from Proto-Saturn are chosen for consistency with the image of "The Eye in the Sky" that our early human ancestors saw in their sky and have passed down to us in their ancient symbols. A thorough discussion of this matter will be presented in Chapter 16. The dimensions of the Sun, Earth, Mars, Proto-Venus, and Proto-Saturn are all drawn to approximately the same scale, but the distances between them are not shown to scale. The diameter of the flux tube is also shown approximately to scale.

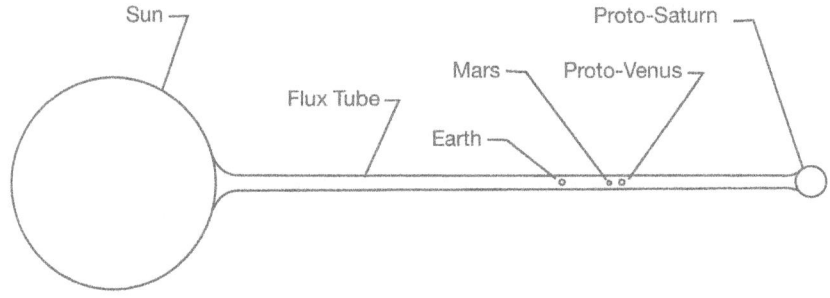

Figure 3: **The Early Solar System**

The Dimensions of the Early Solar System

Figure 3 is drawn with the flux tube at about 33% of the diameter of Proto-Saturn to reflect its calculated size. It is shown with a flare-out (which may or may not exist) where it attaches to the Sun and at Proto-Saturn. Figure 4, below, shows the same features as the above figure, but with all actors and the distances between them shown to the same scale.

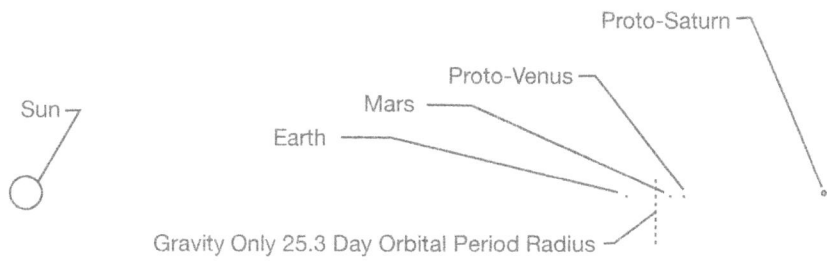

Figure 4: **The Early Solar System Shown to Scale**

In this sketch, the Sun is the size of a pea, Proto-Saturn is the size of a sesame seed, and the largest of the planets is reduced to a size that is slightly smaller than the period at the end of this sentence. Unfortunately, the flux tube is so thin in this rendition that it cannot be shown, but this view does put the aspect ratio (length-to-width ratio) of the flux tube into proper perspective, showing it to be at a very high number, i.e., long and thin. The present-day orbit of Mercury is off-stage to the right, almost twice as far from the Sun as Proto-Saturn is shown, meaning that the orbit of early Earth was about three times closer to the Sun than Mercury is today. Fortunately for us, its inhabitants were protected from the Sun's searing heat by the dusty atmosphere within the flux tube.

The three rocky planets are positioned very close to the gravity-only 25.3-day orbital period radius, shown as a vertical dashed line in the sketch. This is the radius where a planet with that period would orbit freely if the flux tube electrical forces were not there. Planets closer to the Sun than this magic orbit would try to move faster, so they would have a shorter orbital period, and those farther away would try to move slower, so they would have a longer orbital period. This means that Earth is being ever so slightly retarded, and that Mars and Proto-Venus are being ever so slightly advanced from their respective gravity-only orbits by the flux tube electrical forces, resulting in a slight distortion of the flux tube in the vicinity of these planets. (In Chapter 16, titled *The Eye in the Sky*, a more detailed view of this effect will be presented as Figure 10: *Distortion of the Flux Tube by Gravity-Only Orbital Effects*.)

As I was drawing Figure 4, I was intrigued by the close proximity of the three early planets to the gravity-only radius for my selected orbital period of 25.3 days, and by the lack of planets between the Sun and Earth and between Proto-Venus and Proto-Saturn. I knew that plasma physics predicts multiple DL layers at somewhat regular intervals in the plasma contained within a flux tube, and that material tends to collect within these DL points. I asked myself, why are there only three planets in this flux tube when a quick look at the proposed positions of Earth, Mars, and Proto-Venus indicates that there could well have been two or three more planet-containing DL points beyond Proto-Venus, and perhaps as many as twenty between Earth and the Sun?

A truly marvelous "Aha!" moment hit me when I realized that the reason why no other planets formed in any of the several other probable DL points along the length of the flux tube is that these other locations were too far away from the magic gravity-only orbit. Their greater distance from the magic orbit meant that the same retarding or advancing forces mentioned above were strong enough to force any accumulated material out through the wall of the flux tube, allowing that material to either fall back into the Sun or be thrown outward toward Proto-Saturn due to gravitational or centrifugal force unbalance before it could form into a viable planet-sized object.

*Aha! #4:
Proximity to the gravity-only orbit radius
(an orbital mechanics phenomenon)
allowed only 3 planets.*

Now that their location within the solar system has been determined, my next chapter will take a closer look at the probability of life evolving on all three of these molten planets I left hanging in space. At this moment in my story, the stage is set, and the actors are in place. The setting is 1.5 billion years ago (at least). Let the action begin.

Five Friends

Chapter 10

The Evolution of Life

Since the existence of life on Earth is an established fact, let's start this chapter with a discussion of how Conventional Wisdom describes the evolution of life here on our planet and finish with speculation on possible life on Mars and Proto-Venus. For years I have enjoyed a wonderful book that offers a concise summary of the evolution of life on Earth titled *Thread of Life: The Smithsonian Looks at Evolution* by Roger Lewin, published by Smithsonian Books in 1982. I have never had any argument whatsoever with the fossil record described within this reference, or any other scholarly work on evolution, but as this chapter proceeds, I will add my alternate views as a comparison to the evolution-enabling conditions Lewin describes as prevailing on Earth at any particular time. The fossil record of 65-million-year-old dinosaur bones that started me on this journey in the first place is an important part of the story told therein.

The Conventional Wisdom, as presented in *Thread of Life*, proposes the following: Earth formed in the vacuum of space in the form of a thin-crusted molten sphere in its current orbit around the Sun. All the other planets, their moons, and the asteroid belt also formed in their current positions around the Sun. There were several continents on the Earth's surface driven by convection currents within the molten core on which they floated. The Moon, made from material blasted out of the Earth at an earlier time by a collision with an asteroid, was orbiting the Earth. Earth spun on its north-south axis at approximately its current angle to the plane of the ecliptic. The poles were cold. There were seasons.

Five Friends

There was night and day. Once water formed on the surface, there were tides. The atmosphere was gravitationally retained. All constituents of the atmosphere originated within the Earth itself. And Earth shared that atmosphere with no other solar system object.

Compare those conditions to my Unconventional Wisdom's proposed environment: The Earth formed along with Mars and Proto-Venus inside a flux tube that extended from the Sun to its companion star Proto-Saturn. They all orbited the Sun in synchronous rotation with Proto-Saturn, spun on their individual axes at a rotational period of about 24 hours, and had their axes aligned with the flux tube axis. That flux tube was filled with elemental gases that had previously formed in multiple fusion zones in the Sun's corona and compounds that had formed within the flux tube itself, creating an atmosphere that was shared by all three planets.

There was constant light from the Sun and Proto-Saturn, diffused by the flux tube's atmosphere, shining over the entire surface of all three planets. There was no night and day. There were no seasons. There were no tides. Spring-like conditions prevailed continuously. None of the planets had moons.

From this comparison, it can be argued that the Unconventional Wisdom proposes the more favorable set of circumstances for the formation of life. Perhaps the most important of these is a more likely source of the Earth's water. Lewin, in *Thread of Life*, following the Conventional Wisdom, describes gases being "sweated out" of the molten core, surrounding the new planet with an envelope of water vapor, nitrogen, methane, ammonia, carbon dioxide, hydrogen, and other minor components, and that "most of the lighter gases, particularly hydrogen, drifted back into space as the Earth's gravity was too weak to hold them."

Considering that the molten planet undoubtedly stayed molten for hundreds of thousands, if not millions of years before its crust solidified, it is much more likely that the gases Lewin mentioned would have been "sweated out" during Earth's crust-free molten phase. Once released, they would have remained in a high-temperature/low-density state due to heating from the extremely hot planetary surface, and as a result, would have "drifted back into space" just as his previously mentioned "lighter gases" had, thereby leaving the atmosphere void of all the necessary ingredients for the formation of life. Only extremely dense volatiles would have been able to form an atmosphere around a molten Earth floating in free space, much as is the case with Venus today.

In the Unconventional Wisdom I have been developing, this atmosphere retention problem does not exist. In my theory, a large portion of the elements that the Sun had produced in its magnetically confined fusion zones were captured by the flux tube, preventing them from escaping out into free space, and keeping them close by. Once Earth cooled to a low enough temperature, however long that cool-down took, the planet's gravity would attract gases, including those Lewin mentioned as necessary for the formation of life, from within the flux tube, forming a more concentrated local atmosphere near its surface than that which populated the general flux tube volume.

All creatures now living on Earth should be very thankful that the ubiquitous flux tube, which has dominated my discussion up to this point, was in place. Without it, our planet and its two siblings would never have formed. With it, not only did these three planets form, but also water and other necessary life chemicals were made available until conditions were right for life to develop, probably on all three. If it were not for the fact that there were gases trapped along the entire length of the flux tube, the evolving local gravity-concentrated atmospheres around Earth, Mars, and Proto-Venus, all of which were closer to the Sun than Mercury is today, would have been heated to temperatures far in excess of those that were acceptable for the formation of living cells. The dispersion effects of the flux tube's atmosphere attenuated the heat from the Sun to allow life-sustaining conditions to envelop all three planets.

Eventually, the Earth cooled to a low enough temperature for a selective condensation process to commence. The first significant clouds to form consisted of liquid hydrocarbons and oil, which rained down on the planet's surface. Since the Sun-facing side of the planet was slightly warmer than the side facing away from the Sun, most of the oil condensed on the cooler highlands facing Proto-Saturn. The warm liquid flowed downhill, eroding the sloping surfaces over which it ran and carrying the resulting debris along in its flow. As the streams slowed at level areas and low-lying depressions along the way, the debris settled out as oil-rich mud. As the higher volatility elements within the oil evaporated back into the local atmosphere, the lowlands dried up, forming oil-rich sedimentary clays.

After tens of thousands more years, the Earth's atmosphere cooled enough for the life-giving water vapor clouds to appear. By that time, the Earth's surface temperature was only slightly warmer than present-day equatorial temperatures. For hundreds of millions of years, rain fell from these clouds, forming rivulets, streams, and rivers on the barren rocky surface of the Earth. With no vegetation to hold it back, the rushing water eroded the land at a prodigious rate as it flowed inexorably from higher elevations to lower elevations and eventually to the ocean basin on the Sun-facing end of the planet. As it slowed its pace in the relatively flat lands along its route to the sea, and whenever it encountered a natural basin along the way, the water would drop its entrained debris to the bottom of the channel or basin where it collected as mud that would eventually harden into sedimentary rock, much of it burying the previously formed oily sediment.

Since the Sun-facing lowland end of Earth was warmer than the Proto-Saturn-facing highlands, a natural moisture circulation mechanism developed. Water would evaporate from the warm surface of the sea and move inland as vapor where it would condense and fall as rain onto the cooler highland surface below. Most of the moisture fell within a band that extended inland about halfway to the pole. This natural recycling of water, supplemented by the supply of freshly condensed water vapor that was gravitationally attracted from the flux tube's seemingly inexhaustible supply, reshaped the land with the resulting erosion and sedimentation cycle powered by flowing water. Gradually, the lowland filled to become an ocean, and the face of the land changed.

The Evolution of Life

Where the water was shallow along the edges of the Pangaean landmass and in shallow inland seas, conditions were ripe for the formation of life. The warmth of the diffuse light from the Sun, the soup of dissolved minerals in the ocean and inland seas, and the atmosphere of water vapor, nitrogen, methane, ammonia, and carbon dioxide all combined to provide the necessary ingredients for life.

All that was needed was an external energy source to spark the first single-cell life forms into existence.

In *Thread of Life*, Lewin follows the Conventional Wisdom that has ultraviolet light from the Sun providing the energy to transform the primordial soup into the complex organic molecules that eventually acquired the ability to reproduce themselves about 3 billion years ago. To succeed, this approach needs to be supplemented by the dissociation of water vapor to form free oxygen and a resulting atmospheric ozone layer that could protect the emerging life forms from lethal exposure to that same ultraviolet light that gave them life in the first place.

The Unconventional Wisdom that I have been developing is not faced with this conundrum. In much the same way as the Earth's present-day magnetic field, Van Allen Belt, and atmospheric ozone layer protect us from the harsh environment of space, the atmosphere within the flux tube blocked most of the Sun's UV rays, preventing them from reaching the Earth's surface, while the magnetic field of the flux tube prevented lethal cosmic radiation from doing the same. Instead of UV radiation from the Sun providing the energy for complex molecular formation, the life-giving spark was literally provided by the electrically charged nature of the flux tube through the intermediary action of occasional tube-to-Earth lightning bolts.

As lifeforms became more complex, these same lightning bolts provided occasional bursts of radiation that were capable of gene-splitting to induce mutations that eventually allowed ever more complex life forms to evolve.

In the Conventional Wisdom, the formation of single-cell life is said to have first occurred about 3.6 billion years ago, with the formation of the first multicellular life occurring about .6 billion years ago. Perhaps in the exposed environment of space that the Conventional Wisdom describes as surrounding the early Earth, it would have taken that long for that first step toward higher life forms to occur. However, the more benign atmosphere provided by my Unconventional Wisdom's flux tube likely allowed multicellular life to develop in much less time, meaning that single-cell life could have first occurred as recently as one billion years ago.

My evolutionary timeline does not deviate from the last 600 million years of the Conventional Wisdom's timeline, though as I have mentioned, I will on occasion have different opinions as to the causation of certain features of that timeline. From this new vantage point, it is appropriate to take a brief look at the almost one billion years that life has existed on Earth with reference to the indelible fossil record left behind by some of the species, both flora and fauna, that were kind enough to have left their remains in the soft clay of the eroding landscape, thereby preserving their forms for us to see.

The first life forms were single-cell organisms that evolved in the shallow seas at the edges of the vast Pangaean continent. Lewin, in *Thread of Life*, makes the statement that "Around 600 million years ago multicellular organisms appeared, and within a few tens of millions of years, all the basic designs for complex life forms had been established. Since then, life has again and again consisted of variations on a theme." From algae to seaweed, to trilobites, to sponges, to shellfish, to fish—life evolved in the nurturing warmth and light of the early Earth's eternal spring.

About 350 million years ago, a fish developed a flotation bladder that could be filled with a gulp of air and slowly absorbed into its bloodstream, allowing it to venture, however briefly, out of the water and onto dry land. The first amphibian, the lungfish, was born. Animal life was finally able to move onto the land, where, for the moment, there were no predators. Evolution exploded.

The Evolution of Life

Insects evolved and thrived. One dragonfly-like insect evolved to a size that had a 76 cm (30-inch) wingspan. Amphibians evolved into lizards that could eat insects. Plants in the form of grasses moved onto the land and evolved into trees, such as the ginkgo and the tree fern. Height was an advantage for trees, allowing them to more effectively spread their spores into the wind, so they grew very tall. Around 250 million years ago, the first dinosaurs appeared, and within another 50 million years, with their height reaching as high as 15 m (50 feet), allowing them to eat the leaves of the tall trees, they consolidated their reign as the supreme life form on the planet. The Age of the Dinosaurs was now in full swing.

Dinosaurs dominated the land, while their cousins, the sea reptiles and the winged pterosaurs dominated the ocean and the sky. Evolution had shown that size matters. The dinosaurs were the biggest, heaviest, strongest species that had ever lived on the planet. The land they lived on was a lush, verdant Eden with perpetual light, gentle rain, a viable atmosphere, and moderate temperatures.

Life for the ancestors of the menagerie that shared the Age of the Dinosaurs had not always been easy. Species did not thrive forever. Extinctions occurred. Variations in carbon dioxide, oxygen, and methane levels, chemicals from toxic rain, and even flood water in the form of rain (all of which occasionally migrated into the Earth's atmosphere from the flux tube's substantial atmosphere) are possible causes for these extinctions. Temperature changes up or down could have resulted from changes in the opacity of the flux tube gases, or from changes in the distance between the Sun and the Earth as a result of slight changes in the flux tube's voltage gradient. All of these extinction-producing calamities occurred from time to time on the flux-tube-protected ellipsoidal Earth.

Of the twenty or so significant extinctions that stand out from the general background extinction rate, there are five major events that are usually highlighted in the available literature on evolution. The first occurred about 444 million years ago, the second occurred over the timespan of 407 to 359 million years ago, the third from 266 to 251 million years ago, and the fourth around 200 million years ago. In each of these, between 70%

and 95% of all marine or land species living at the time became extinct. During the Age of the Dinosaurs, the fifth of these major extinctions had not yet happened.

A typical species may have had a 10-million-year life span from its first appearance to its disappearance. Some species had a continuous line of evolution from the early fossil record to the present day. Some made a slow, gradual appearance, and then faded away. Some burst onto the scene. Some suddenly disappeared. Charles Darwin's concept of "the survival of the fittest," or perhaps better stated as "the survival of the most adaptable," is the usual explanation for the gradual appearance and disappearance of a species. However, in the case of the mass extinctions that periodically interrupt the entire span of time that life has existed on Earth, "the survival of the luckiest" is perhaps a more appropriate phrase.

So far, I have devoted most of this chapter's discussion to life on Earth. The vast first-hand evidence of Earth's rich fossil heritage allows the relative timing and extent of evolution on our planet to be well documented. But what about the other two planets, Mars and Proto-Venus, that formed at the same time as Earth and were cocooned inside the same flux tube as Earth, sharing the same atmosphere? Did life form there too, and if it did, does it still exist today?

The answer to this question lies primarily in the similarities and differences between the three planets. Mars was about 47% smaller in diameter than Earth, and Proto-Venus was about 10% larger in diameter than Earth. All three planets shared the same flux tube atmospheric conditions, though there were probably slight differences based on their relative distance from the Sun due to solar radiation heating.

As I discussed in the previous chapter, "…the orbit of the early Earth was about three times closer to the Sun than Mercury is today." Were it not for the shadow of the atmosphere within the flux tube, the surface temperature on all three planets would be too high for life as we know it to exist. But since the flux tube was surrounding each planet with an atmosphere that would provide all the elements necessary for life, and

The Evolution of Life

the cool shade of its shadow, the only absolutely essential requirement for life as we know it to evolve on any of the three nascent planets was water. No water, no life! But there was another influence in play that must be considered, namely: on which side of each planet did its primordial ocean basin form.

At the time the three rocky planets formed, there was a subtle phenomenon that affected each of them depending on their location relative to the gravity-only orbital radius (which I will define for you momentarily) and the Sun. In the previous chapter, I provided a description of how electrical forces in the flux tube altered the orbit of Proto-Saturn, provided the force that distorted and reduced the effective gravity on the three rocky planets, and kept all three planets and Proto-Saturn aligned in a synchronous orbit around the Sun. These electrical forces acted on voltage differences alone and were completely independent of the mass or material composition of the planets themselves. At the same time as these electrical forces were acting, the Sun's gravitational force and circular-motion induced centrifugal force also acted on all three planets, except these forces were only a function of the mass of the planets and the material thereon, and were completely independent of the local electrical charge.

The gravity-only orbital radius is defined as the only distance from the Sun where the Sun's gravitational force is perfectly balanced with a planet's centrifugal force for a given orbital period of that planet around the Sun. If an object with the same orbital period is closer to the Sun, the Sun's gravitational force dominates. If that object is beyond that radius, centrifugal force dominates. Figure 4 in the previous chapter shows the Earth in a position that is closer to the Sun than the gravity-only orbital radius, with Mars and Proto-Venus in positions that are beyond that radius.

As the planets cooled, this difference in position created a difference in the segregation and migration of material floating on each planet's molten core. The heaviest of the mineral compounds, those with the greatest mass density, such as basalt, would tend to move in the direction of the strongest force, and the lighter, less-dense compounds, such as granite, would be displaced to move in the opposite direction. This is the same effect that causes heavier liquids or finely dispersed, higher-density solid particles to settle to the bottom of a liquid-filled container when left undisturbed in a gravity field. Centrifuges that spin such containers at high speeds enhance this effect by creating a radial centrifugal force many times stronger than gravity.

Since heavier basalt tends to form ocean floors and lighter granite tends to form continents, it is easy to predict the side on which a primordial ocean basin would be expected to form for each of the three planets. For planet Earth, the expected side would be the side facing toward the Sun, where gravity dominates, while a predominantly granite primordial landmass would be expected to form on the side facing away from the Sun. This is indeed the way I have described the formation of the Earth's surface earlier in this chapter. For Mars and Proto-Venus, however, just the opposite would be expected, with ocean basins forming on the side away from the Sun, where centrifugal force dominates, and their individual primordial landmasses forming on the Sun-facing side.

There is evidence at hand that, at least for Earth and Mars, this is indeed what happened. For reasons that will be revealed in detail later in this book, there is no way of knowing whether these expectations were also true for Proto-Venus, but for now, I will assume that they were. Suffice it to say, this primordial continent-orientation phenomenon is a pivotal "Aha!" moment for my thesis and provides a convincing argument for its validity. Thus,

---

Aha! #5:
Primordial Continent-Orientation (PCO) matters.

Let me get back to the subject at hand: namely, did life form on Mars and Proto-Venus? From the preceding presentation, it would be expected that the location of the primordial continental landmass would have an effect on the formation of life, giving Earth the advantage over Mars and Proto-Venus. I'll start with Earth. When our own rocky, ellipsoidal planet was just a molten mass of magma with a liquid metallic core, the

The Evolution of Life

predominance of the Sun's gravity over centrifugal force positioned 60% of the Earth's surface as dense basalt lowlands on the warmer, Sun-facing side of the planet. The remaining 40% of the surface, composed of less-dense granite that tended to be thicker and float higher on the Earth's molten mantle, formed the highlands known as Pangaea on Earth's cooler side.

As might be expected, the difference in density between these two zones created a slight difference in surface elevation between the two ends of the planet. The Sun-facing 60% of the planet was at a lower elevation than the remaining 40% facing Proto-Saturn by only about 8 km (5 miles), which is just 0.12 % of the Earth's average radius. As the Earth cooled over a period of millions of years, the surface of the magma hardened, resulting in a solid shell only 97 km to 194 km (60 to 120 miles) thick, or about 2% of the 6,400 km (4,000 miles) average radius core on which it floated. The 8 km (5 mile) elevation difference was solidified in place and though very small, was enough to allow the lowlands facing the Sun to act as a basin for the eventual collection of the water that formed the Earth's oceans.

Mars, too, was just a mass of molten magma with a dense molten core, but in its position just beyond the gravity-only orbital radius, the predominance of orbital centrifugal force over the Sun's gravity acted to swap the position of the planet's lowlands and highlands. As a result, the more dense lowlands, known today as the North Polar Basin or Borealis basin and covering about 40% of the planet's surface in what is now its northern hemisphere, wound up on the cool side of Mars, facing away from the Sun. The remaining 60% of the surface, consisting of the planet's highlands, ended up facing the Sun on the warm side of Mars.

As Mars cooled and the surface of the magma hardened to become the planet's crust, this slight difference in elevation, about 1 to 3 km (.62 to 1.7 miles), between the two ends of the planet, with an average thickness of 32 km (19.9 miles) in the northern lowlands and 58 km (36 miles) in the southern highlands, was frozen in place, forming what is now the planet's most conspicuous feature. This sharp contrast in elevation between the two hemispheres is known today as the "Martian dichotomy."

Earth's lowlands, where its primordial ocean formed, were on the warmer Sun-facing side of the planet, whereas the lowlands of both Mars and Proto-Venus, where it is very likely that a primordial ocean also formed, were on the cooler side of the planet facing Proto-Saturn. For Earth, this arrangement provided solar heating on the Sun-facing side, allowing water to evaporate, enter the local atmosphere, and then condense as rain on the cool Proto-Saturn-facing highlands where it would water the ground, allowing plant growth, with the run-off collecting in rivers to flow back to the warm ocean. This is the perfect setup for allowing life to flourish on the land.

For Mars and Venus, with their oceans on the cool Proto-Saturn-facing side, this life-giving, heat-driven water conveyor belt did not exist. The Sun-facing highlands would get hot, causing any moisture to evaporate into the local atmosphere, where it would condense as rain over the cooler oceans, preventing moisture from returning to the dry land.

There is every indication that Mars had a primordial ocean on its Proto-Saturn-facing side in which aquatic life could have evolved, perhaps even some land animals and plants along its immediate coastline. The present-day surface of Mars reveals many clues that it once had an ocean covering its lowlands and that massive amounts of flood water once flowed from the highlands near the shoreline back into its ocean. However, with none of the long river valleys winding from the heart of the highlands flowing back to the sea as we see on Earth today, the likely existence of a robust land animal population is highly unlikely. Chapter 19 will reveal the source of the aforementioned Martian floodwaters and how Mars subsequently lost its ocean and atmosphere.

There is every reason to believe that Proto-Venus also had an ocean on its Proto-Saturn-facing side where life could have evolved, but the present surface of Venus holds no clue as to the actual surface of Proto-Venus. Please be patient until Chapter 19 when all will be revealed.

As I end this part of my story, back on Earth the Age of the Dinosaurs was thriving, and the world was in full bloom. The geological time was 65 million years ago.

Chapter 11

The Extinction of the Dinosaurs

Ten chapters ago, I introduced my five imaginary friends, one of whom was a geophysicist who was quite proud of his branch of science and its discovery of "… an asteroid crater in Mexico's Yucatán Peninsula that provides the smoking gun for the dinosaur extinction." I then used ten chapters to develop an explanation for how the dinosaurs managed to exist in the first place. In this chapter, I will explain the actual mechanism that caused the extinction of those magnificent creatures.

There is no need for an external event such as an asteroid impact to disrupt the dinosaurs' Eden-like world. All that is required is a change to the electrical forces responsible for reducing Earth's effective gravity, which made it possible for the dinosaurs to stand. Throw in a global shape change as the Earth slowly adjusted to its new effective surface gravity, and you will have the inescapable cause of the mass extinction that destroyed the dinosaurs and 80% of all animal life on Earth 65 million years ago.

It is appropriate to begin this part of my story by taking a detailed look at our home planet as it existed just before the dramatic events that literally changed the world. Figure 5 below provides three views of the Earth as it would have appeared 65 million years ago:

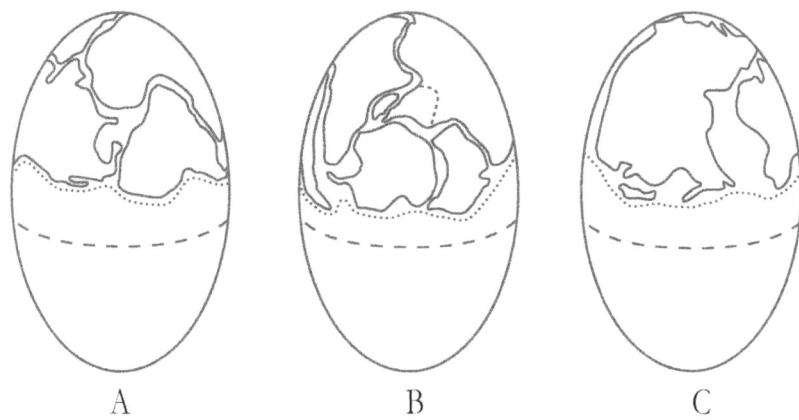

Figure 5: **The Ellipsoidal Earth, 65 million years ago**

These figures present three different views of the Earth and its primordial continental landmass, Pangaea, as my theory predicts their appearance would have been 65 million years ago. The modern-day continents are shown in a solid outline in their familiar shapes, approximately as their borders would be seen on a present-day globe. View A shows North America and a small corner of Eurasia on the upper left, and South America and part of Africa on the upper right. View B shows parts of Africa and South America on the upper left, the subcontinent of India in contact with part of Eurasia at the dashed line on the upper right, and Antarctica and Australia in the center. View C shows Eurasia on the upper left and part of North America on the upper right.

The vantage point for these views is about 60,000 km (37,000 miles) above the Earth's surface and about 60,000 km (10,000 miles) above the plane of the equator, with the Sun about 2.38×10^7 km below the Earth and Proto-Saturn about 0.8×10^7 km above the Earth (the flux tube is not shown so that the details of the Earth's surface can be seen). The continents are shown touching their neighboring continents at their approximate continental shelf contours so that they form Pangaea's one continuous landmass. The equator is shown as a dashed line and appears as an elliptical segment due to the elevated viewing angle. The dotted line between the individual continents and the equator represents the approximate contour of each continent's shelf on the side that does not contact another continent. The primordial ocean occupies the entire bottom half of the Earth up to the dotted continental shelf line. These three views show how the planet looked during the entire time that life was evolving on Earth up to the Age of the Dinosaurs.

The Earth I have shown in these views had a local atmosphere enveloping the planet that provided atmospheric pressure not significantly different from what we experience today. But unlike today, that atmosphere was intimately connected to the atmosphere trapped inside the entire length of the flux tube and was shared by the other planets therein. The air immediately around the Earth was at a higher pressure than the flux tube gasses due to the gravitational attraction of the Earth's mass. Over time, the composition of the atmosphere in the solar system's flux tube would change due to the circulation of gasses and differences in elemental emissions from the Sun. As a result, the composition of the Earth's local atmosphere would occasionally fluctuate as it maintained its equilibrium with the vapor pressure of the flux tube gasses. Since water vapor was a significant constituent of that local flux tube atmosphere, the amount of water available to fall as rain would also fluctuate, causing changes in the Earth's ocean level.

If the ocean level was low, the entire upper half of the planet, which we now identify as Pangaea, would be gently rolling dry land punctuated by small, shallow seas in various low spots and carved by rivers draining the higher elevations to these inland seas and eventually to the ocean itself. Under these conditions, land animals had the run of the entire primordial continent. If the ocean level were to increase, more and more of the continental shelf land would flood, providing more shallow-water area for aquatic animals to exploit, while occasionally isolating portions of the planet and the land animals living thereon. This variability in available habitat was a major contributor to the evolution and extinction of species, both aquatic and terrestrial, during the time of the Early Solar System.

There are some interesting implications regarding the dinosaurs' habitat that can be deduced from the images shown in Figure 5. I did some research to determine the locations on the Earth's surface where coal beds were prominent and where, coincidentally, most dinosaur fossil finds are located. That study revealed that the United States on the North American continent, Russia and China on the Eurasian continent, the Australian continent, and the Indian subcontinent account for more than 65% of coal reserves in the world today, and that the South American continent and the southern part of the African continent also have significant quantities of coal. Even the Antarctic continent has coal discovered in 1912 but does not make the list of coal producers since its international treaty provisions prevent any mineral exploitation.

As can be seen in Figure 5, all of these coal producers lie closer to Pangaea's shore than to its center at the pole. On the other hand, the area around the center of Pangaea, which was located near the present-day eastern end of the Mediterranean Sea, has very little, if any, coal reserves. This distribution of fossilized carbon that formed the coal beds provides mute testimony to the preference of the early Earth's flora for the land areas closest to the primordial oceans, indicating that ocean moisture did not reach the polar areas in sufficient quantities to produce the lush forests that the plant-eating dinosaurs relied on for sustenance. These geological and geographical facts merit an Aha! moment of their own—

Aha! # 6:
The location of coal deposits is compatible with a hemispherical Pangaea at one pole of an elliptical Earth.

In earlier chapters, while discussing the characteristics of an ellipsoidal Earth, I described the reasons why the effective gravity thereon would not be uniform. The combined effect of gravity and the flux tube's axial and radial electrical forces could be proportioned to produce a 0.33 G field at the equator, but the effective G at the poles would necessarily be less in order to preserve the planet's ellipsoidal shape.

I also described the first clue used to understand the existence of the dinosaurs, namely the fossilized remains of Argentinosaurus, which, as its name implies, was found in Argentina, very close to the early Earth's equator (see Figure 5: View A). The location of present-day coal deposits provides insight as to why the largest of the dinosaur species did not populate the even lower G-level areas near the more sparsely foliated center of Pangaea. They were plant eaters that had evolved where the vegetation was most lush. The ones with the longest necks grew to be the biggest since they could graze in the treetop canopy where the leaves of the giant ginkgo and tree-fern plants thrived on the moisture from the nearby sea.

Up until 65 million years ago, ocean level changes and atmospheric composition changes were the primary causes of the four major extinctions and the general background extinction level that life on Earth experienced. But there was something significantly different about the fifth major extinction that occurred 65 million years ago. The dinosaurs, pterosaurs, and sea reptiles were completely wiped out. Never before had an entire group of animals been cut off with no surviving members left to carry on or to gradually die out. The strongest clue as to what caused this massive extinction is that no land animal with an adult weight of more than 23 kg (50 pounds) survived. Mammals, land turtles, snakes, lizards, birds, and crocodiles that met this weight criterion survived and continued to diversify, but only the small survived.

The dinosaurs and the pterosaurs share a common trait in that they personified the successful evolutionary trend that "bigger is better." They were huge creatures that had evolved in the low-gravity environment of the early Earth. From my earlier discussion of the dinosaur/gravity problem in Chapter 3, it is obvious that if the Earth's effective gravity were to double, the dinosaurs would not be able to support their weight with a skeleton and muscles that had evolved in a low-G environment. Similarly, the pterosaurs would no longer be able to fly with a wing area to body weight ratio that could only allow flight in a low-G environment.

The sea reptiles were air-breathing animals that had adapted to the ocean and needed the water environment for survival. They too had evolved to extremely large sizes. The largest sea reptile, the liopleurodon from the pliosaur family, weighed perhaps as much as 150,000 kg (150 long tons), making it heavier than even the largest of the dinosaurs. Thanks to Archimedes' Principle, which states that "a body displaces its own weight in a liquid," their buoyancy supported their massive weight and made them essentially independent of gravity. They would not have been affected by a doubling of Earth's effective gravity since both the weight of the animal and the volume of water it displaced would increase by the same amount, yet they still perished along with the dinosaurs. This means that there had to be an additional extinction mechanism, other than an increase in gravity, that was in play during the dinosaur extinction event, and that mechanism had to be more deadly to air-breathing sea reptiles than to fish, birds and air-breathing land animals.

One such mechanism would be poisonous gas from volcanic eruptions. On land, those toxic gases would be hot and would tend to rise into the atmosphere, where they could disperse before their concentrations became lethal. However, in an undersea eruption, the poisonous gases would cool to the surrounding water temperature as they bubbled to the surface, causing the denser toxic gases to concentrate at the water's surface until they were dispersed by winds into the greater atmosphere. Such a scenario would allow fish that breathe underwater through their gills, as well as land animals and low-flying or ground-resting birds, to escape the highly concentrated toxic gas. However, it would immediately poison any air-breathing sea animal as it surfaced to breathe and instead inhaled the concentrated toxic gases that collected just above the water's surface.

So how then did the Earth's environment change to kill off the dinosaurs on land, the pterosaurs in the air, and the reptiles of the sea while sparing small land animals, birds, and fish? The answer—electricity! The ellipsoidal Earth, with a gravity-attenuating electrical field that I have proposed, has all the features necessary to provide an increase in surface gravity along with the volcanic eruptions just described.

Here is the proposed scenario: Ever since the stellar fission that gave birth to Proto-Saturn, the flux tube, and three rocky planets, the Sun had

The Extinction of the Dinosaurs

experienced only slight voltage variations during the entire time that life developed on Earth. The only consequence of these variations was slight changes in the strength of the electric field between Proto-Saturn and the Sun, resulting in small variations in the distance between them, and between the three small egg-shaped planets trapped in the flux tube. The planets themselves would have experienced only slight changes in effective surface gravity with each of these voltage changes, perhaps leading to weak earthquakes or periodic volcanic venting.

But sometime around 65 million years ago, the Sun gradually moved into a region of the Milky Way galaxy where the ambient electrical charge was substantially less than what the Sun and its solar system had experienced up until that time. The resulting decreased voltage difference between the Sun and Proto-Saturn caused the axial and radial forces in the flux tube to slowly decrease and to do so at different rates. The reduced axial force allowed Proto-Saturn to pull further away from the Sun, causing the flux tube to become severely weakened as it was strained, but it was not destroyed. The three planets trapped within the flux tube slowly felt the same significant decrease in the axial force that had stretched them into their ellipsoidal shapes when they first formed, but since the axial force dropped more than the radial force, the planets were forced to change shape. Each planet saw its pole-to-pole distance shrink, its equatorial diameter increase, and felt its effective surface gravity almost double. Mars, the smallest of the three, was the least affected by these changes. Proto-Venus, the largest, felt the effects even more severely than Earth but managed to just barely survive.

On Earth, the resulting gravity and shape changes were catastrophic for the animals living there. The gravity change alone sealed the fate of the multi-ton dinosaurs and the large flying reptiles, the pterosaurs. If the voltage drop had been sudden, the dinosaurs would have collapsed almost immediately under their own weight, and the pterosaurs would have fallen out of the sky. However, several factors point to a gradual increase in the Earth's effective gravity that took about 30,000 years to occur.

Despite the slow nature of this change, it was not slow enough for evolution to keep up. Over successive generations, the dinosaurs became heavier and less capable of functioning, while the large pterosaurs, after losing the ability to take off from the land, eventually lost their ability to function in their grounded state. Any animal species weighing more than 23 kg (50 pounds) was at a competitive disadvantage and eventually lost out to species better suited to the increased gravity environment. But if gravity change had been the only effect of the Sun's voltage change, the resulting extinction event would have been much less devastating.

According to Conventional Wisdom, the Earth of 65 million years ago had the continents arranged on a spherical surface, with gravity identical to what we feel today. That wisdom, though it does not explain the existence of the dinosaurs, explains their extinction as a result of a collision between the Earth and a large asteroid. The resulting shockwave is said to have swept around the globe, spreading debris and dust that blocked the Sun's light and warmth, causing an "asteroid winter" that lasted for years. Plants died, and the plant-eating animals that fed on them perished. The meat-eaters soon followed them into extinction. Numerous other theories exist, but this is one of the current favorites.

The strongest evidence supporting this scenario is a thin layer of clay rich in iridium and coesite (shocked quartz) found widely distributed over the surface of the globe at a depth coincident with the time of the dinosaurs' extinction, and the discovery of a crater near the town of Chicxulub in what is now the Yucatán Peninsula in southern Mexico, dated to the same period. Since iridium is an element that is rare on the surface of the Earth but would be found in high concentrations in asteroids, a collision with an asteroid would be expected to generate a cloud of iridium-rich dust. The high pressure from the collision would also be expected to produce the high temperatures and pressure necessary to form the coesite found with the iridium.

Conventional Wisdom's proposal that the solar system has been arranged essentially as it is today since its formation, and that the asteroid belt is primordial material that never quite formed a planet due to gravitational disturbances from Jupiter, makes a rogue asteroid collision with Earth a distinct possibility.

My Unconventional Wisdom hypothesis, however, proposes that the asteroid belt did not exist at the time of the dinosaur extinction, making an asteroid collision unlikely. Instead, I propose that the iridium and coesite layer used to define the sedimentary layer of that extinction was deposited by massive and continuous volcanic eruptions that spewed lava rich in iridium and coesite from at least 70 km beneath the iridium-poor crustal rock. These eruptions lasted for 30,000 years due to the fracturing of the Earth's crust as it slowly changed shape from one ellipsoidal form to a less exaggerated one.

There is no denying that the Chicxulub crater exists, but it is not the "smoking gun for the dinosaur extinction" as my geophysicist friend insisted at the beginning of my story. Instead, think of it as the result of a taser blast from a flux-tube-to-Earth thunderbolt. All the signs point to this alternative explanation.

The Chicxulub crater, about 150 km (93 miles) in diameter and 20 km (12 miles) deep, has a peak ring at its center. Such a crater would form if a spark was initiated from Earth to the flux tube, triggered by a process known as "fracto-emission" that occurred as rock particles broke up in response to surface fractures from the Earth's slow-motion shape change described earlier. (Note: a crust fracture passes close to the town of Chicxulub in what is now the Yucatán Peninsula, as shown in *Figure 7: The Earth's Shape Change from 65 Million to 34,000 Years Ago*.)

Once the conductive path from the flux tube to the Earth's surface was established, a powerful thunderbolt would form, carrying its electrical current on its outer surface. As it struck, the energy from the bolt would vaporize a ring-shaped scar on the Earth's surface. The center of the ring, though not vaporized, would be subjected to intense shockwaves that would melt rocks and shock quartz, leaving a ring-shaped crater with an island at its center.

This finding merits the notation of another Aha! moment:

Aha! #7:
The Chicxulub crater is consistent with a flux tube-to-Earth thunderbolt initiated by the generation of a charge through the breakup of rock particles —fracto-emission.

Chapter 12

Continental Drift

In Chapter 2, "The Conventional Wisdom," only a sentence or two appeared on the subject of continents and mountain building. I would like to briefly revisit that subject to show how my new approach to the Earth's mountain-building mechanism and the breakup of Pangaea provides a more feasible explanation than the Conventional Wisdom and its premise of a spherical primordial Earth.

The founders of Conventional Wisdom examined the spherical Earth of today and observed that the entire crust is made up of fractured tectonic plates, some of which have a continent embedded in them and others that do not. They noticed that these plates were moving and could possibly have fit together like a giant jigsaw puzzle sometime in the distant past, with all the continents touching each other to form a single irregular-shaped landmass they named Pangaea.

Mountain ranges that seem very ancient, located well within the borders of the modern continents, were observed, and they concluded that just as the continents have drifted apart since the time of Pangaea, even earlier continents must have collided to form these mountains in the distant past. They also noted some very rugged (obviously more recent) mountains with jagged ridges and peaks, distinguishing them from the more rounded, smooth-ridged older mountains. They concluded that these newer mountains were similarly formed when the broken pieces of Pangaea collided with each other in more recent times. Once they decided that continents could drift and indeed had been drifting around on a spherical ball of molten lava since continent-sized pieces of crust first solidified, they proposed that convection currents in the molten core flowing under the Earth's solid crust were the motivating force causing the fracturing, drifting, and mountain building that ensued.

Unfortunately for the Conventional Wisdom, whenever I observe how freely molten lava flows during a volcanic eruption, it is hard for me to imagine that the viscosity of lava, driven by convection currents, is sufficient to the task of rending the Earth's crust and building mountains. On the other hand, if an ellipsoidal Earth shifted its shape as I have been proposing in my Unconventional Wisdom, the dramatically stronger force of gravity would be available to fracture and compress the Earth's crust.

I, too, was thrilled when plate tectonics revolutionized the geosciences. I am a very geometry-oriented engineer who instinctively knew that continental drift, whatever it was called, was real—but the notion that the Earth's crust could be fractured by the viscous drag of convection currents in its molten core just didn't add up. Watching lava flows from Hawaiian volcanic eruptions is sufficient to demonstrate that the viscosity is too low to fracture granite and basalt rock when it can't even tear pieces of its own hardened shell from the lava tubes it forms. I can accept that it is sufficient to cause subduction of basalt sea floors where there is already a fracture, but not strong enough to cause the fracture itself. The shape-shifting Earth I propose has the muscle to build mountains and fracture the solid primordial crust of Pangaea into continent-sized pieces that can subsequently be pushed around by convection currents.

The Conventional Wisdom's problem starts with the belief that the Earth has always been spherical. What if the Earth that formed inside the flux tube of my earlier discussions had been spherical instead of ellipsoidal, and what if, at the same time, I could ignore the fact that without gravity gradient stabilization, it would quickly lose its spin axis orientation with the Sun (see Chapter 4). How would such a spherical Earth have responded to the voltage change I proposed as the cause of the ellipsoidal Earth's shape-shifting?

Suppose the axial and radial electrical forces inside the flux tube proposed for my Unconventional Wisdom Earth were to have been proportioned such that the 1/3rd G Earth of the Early Earth had assumed a spherical shape. And suppose that those forces remained proportioned such that the Earth retained its spherical shape after the electrical forces dropped dramatically at the end of the Early Earth. In this scenario, the 1/3rd gravity necessary for the existence of the dinosaurs and the airborne pterosaurs would have increased to the point that they would have been unable to survive. With the shape of the Earth being spherical both before and after the gravity increase, there would have been virtually no effect on the Earth's crust. The spherical shape would not change, and the surface area would have been the same both before and after the G change. In short, "Where's the beef?" There is no driving force for plate tectonics other than the Conventional Wisdom's convection currents in the Earth's molten mantle.

My Unconventional Wisdom uses the planet's unimaginably powerful shape-shifting mechanism to pull on the crust in one direction while crushing it in another. The events ascribed to plate collision in the buildup of Pangaea by the proponents of the Conventional Wisdom are instead the result of extensive crust folding caused by the pole-to-pole circumferential contractions at the end of the Early Earth.

It is appropriate at this point to take a close look at the Conventional Wisdom's map of Pangaea on a spherical Earth and review the implications of the continental movements occurring after Pangaea broke up, as proposed therein. Figure 6, presented on the following page, shows one version of the Conventional Wisdom's theory of continental drift.

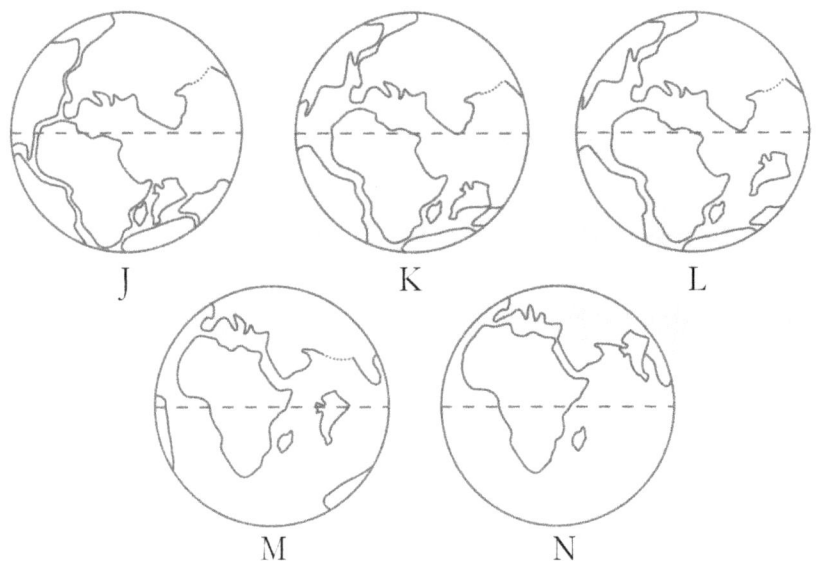

Figure 6: **Continental Drift – The Conventional Wisdom**

The views are from a vantage point that is about 64,000 km (40,000 miles) above a spherical Earth's surface and centered directly over the equatorial region of Africa. View J shows Pangaea 225 million years ago with the Tethys Sea straddling the equator, North America and Eurasia above the equator, and South America, Africa, India, Antarctica, and Australia below the equator. The primordial Pacific Ocean is out of view on the far side of the globe and contains no other continents. View K shows the world at the beginning of the post-Pangaean continental separation about 200 million years ago. View L shows further separation between the continents about 150 million years ago. View M shows even more separation about 65 million years ago, at the time of the extinction of the dinosaurs. View N shows the continents in their present-day positions.

The framers of the Conventional Wisdom had a particular problem with India. They knew that India is currently a subcontinent firmly attached to Eurasia at the foothills of the Himalayas. They also knew that in 1912,

Continental Drift

the geophysicist and meteorologist Alfred Lothar Wegener (1880–1930) observed that certain plant and animal fossils could be found in specific areas of South America, Africa, India, Antarctica, and Australia, giving irrefutable credence to the theory that these five continents all touched each other at the time of the supercontinent Pangaea. As I review the sequence of continental movements from View J through View N, pay particular attention to the continent of India and its position with respect to the broken-line border portion of the Eurasian continent to see how these framers resolved their conundrum.

In View J, India is clustered with its southern hemisphere neighbors, South America, Africa, Antarctica, and Australia, just as Wegener's fossil map would have predicted. In View K, it has pulled away from the Horn of Africa, along with Antarctica and Australia. In View L, it has broken away from Antarctica and Australia and is moving upward toward the equator, all alone, leaving Africa, Antarctica, and Australia behind. In View M, it is crossing the equator while Antarctica has moved out of view to the south, and Australia has moved eastward, almost to its current position. In View N, India has collided with the broken-line border portion of Eurasia, creating the Himalayan Mountains in the process.

Note that in Views K, L, and M, the Tethys Sea is closing as Africa and Eurasia rotate toward each other about a pivot point near what is now the Strait of Gibraltar. In those same views, all the other continents are moving away from each other, except for India, which seems to be riding a conveyor belt headed for its collision with Eurasia. India is the only continent that has collided with another and attached itself thereto, and if it weren't for the Wegener fossil evidence, the Conventional Wisdom's Pangaean breakup sequence would undoubtedly show India attached to Eurasia in all five of these views. This strange movement of India is not required in the Unconventional Wisdom of my shape-shifting ellipsoidal Earth thesis.

Earlier in this chapter, I discussed some of the problems my Unconventional Wisdom has with a spherical Earth, and back in Chapter 10, I described the effect of gravity and centrifugal forces on the ultimate location of ocean basins. With those two discussions in mind, it is appropriate for me to mention another observation I have made regarding the Conventional Wisdom's explanation of the Earth's formation. If, as that wisdom states, the Earth formed in the vacuum of space in its current orbit around the Sun as a molten sphere spinning on an axis that was more or less perpendicular to the plane of the ecliptic, the Earth's ocean basin would be expected to form in a very specific location.

Metallic and other high-density compounds would sink to the core of the molten sphere, while lower-density materials would tend to float on top. The predominant floating rocky materials that formed the surface of this liquid ball would be basalt and granite, with basalt being the denser of the two. Just as a centrifuge tends to move denser material away from its spin axis, a spinning molten Earth would tend to cause the denser basalt material to migrate toward the equator, while the less dense granite material would tend to migrate toward the poles. Since basalt is Earth's predominant sea-floor material and granite is its predominant continental landmass material, once the planet cooled and water filled the lowlands, the resulting Earth would be configured with two primordial continents, one at the North Pole and one at the South Pole, with a single ocean girding the Earth at its equator.

I have made a few simple calculations to determine the size of these polar landmasses and find that the North Pole continent would extend to roughly 35 degrees north latitude, or about where the North Carolina/South Carolina border is located. And as might be expected, the corresponding South Pole continent would extend to roughly 35 degrees south latitude, or about the location of Santiago, Chile. Such an Earth would be very amicable to the formation of life, with a warm equatorial ocean pumping water vapor into the air and cool northern and southern continents where rain could fall in the temperate coastal areas and snow could fall at the poles. It would have also been a very stable configuration for the Earth's landmasses, one that would have been expected to survive to the present day. However, it is apparent that this did not happen. But I digress.

Chapter 13

Mountain Building

Paul Bunyan had such a "mountain-sized" stature and strength that legendary imagery pictures him along with his giant blue ox, Babe. Here, imagery can aid in understanding how the unimaginably strong crustal forces in our shape-shifting planet resulted in the building of Earth's mountains—Figure 7: *The Earth's Shape Change from 65 Million to 34,000 Years Ago*. On the following page, you will see three views each of the Earth at three critical moments in its history.

The first row of Figure 7 is a copy of Views A, B, and C from Figure 5, previously discussed in Chapter 11, showing Pangaea and the original location of the present-day continents as the world appeared from the time the crust first solidified until the end of the Early Earth, 65 million years ago. Its ellipsoidal shape had a length-to-diameter ratio of approximately 1.6 to 1 throughout that entire time period.

The second row, with Views D, E, and F, represents the Earth in its second ellipsoidal incarnation just after the Earth spent about 30,000 years shifting from a 1.6 :1 ellipsoidal shape to a 1.4 :1 ellipsoidal shape, while the fifth major extinction event (ending the Age of the Dinosaurs) was occurring. This was the Earth's configuration during the period I will call the Interim Earth, which stretched from the start of the fifth major extinction up until about 34,000 years ago. During this shape-shifting action, the Earth's volume did not change, but its surface area was reduced by about 2.2%. All that extra surface area had to go somewhere, and the answer to where it went is "mountain building."

The third row, with Views G, H, and I, shows the Earth just before the end of the Interim Earth with the same ellipsoidal shape as in Views D, E, and F, but after almost 65 million years of continental drift had occurred.

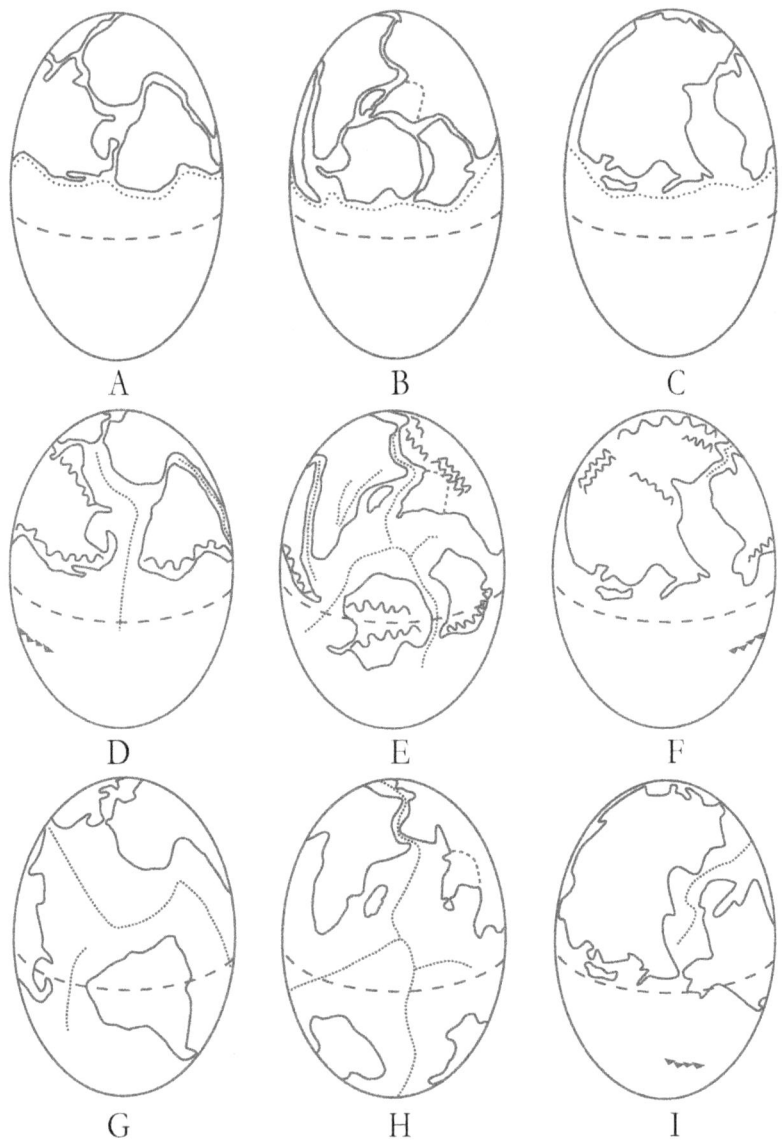

Figure 7 : **The Earth's Shape Change from 65 Million to 34,000 Years Ago**

As you view Figure 7, keep in mind that the sphere is the most efficient geometric shape in terms of surface area. It has the lowest surface-to-volume ratio of any solid geometric form. The ellipsoid (prolate spheroid) is fairly efficient, but all ellipsoidal shapes will have more surface area for a given volume than the sphere. As a volume-constant ellipsoid moves closer to a spherical shape, its pole-to-pole distance decreases, and its equator circumference increases. This means that as the surface area of our volume-constant planet decreased, it did not do so uniformly. During the fifth mass extinction event, the circumference around the Earth's poles reduced by about 7%, while the equatorial circumference increased by about 6%. This means that virtually all of the Earth's land surfaces and sea floors were subjected to a combination of tension in one direction and compression in a direction perpendicular to that tension.

Let's look at the effects of this combined tension and compression loading by considering this three-example description: In the first example, imagine a square sheet of paper with tension pulling uniformly on all four edges. I'll call this T & T loading. The paper will remain flat with tension forces in all directions until the load increases to a level that exceeds the tear-resistance strength of the paper, at which time it will rip apart at its weakest point.

In the second example, imagine that same square sheet of paper with compression *pushing* uniformly on all four edges. I'll call this C & C loading. As the compression load is increased, the paper will start to buckle and fold in a random pattern at a load that is several orders of magnitude lower than the load that tore the paper in the previous example. As the load is continuously applied, the paper will eventually be squeezed into a crumpled wad.

In the third example, imagine a new sheet of paper that is the same size as in the previous two examples, but this time with tension applied to two opposing edges and compression applied to the other two opposing edges. I'll call this T & C loading.

At a very low load, parallel U-shaped ripples will appear in the paper aligned with the direction of the tension force and perpendicular to that of the compression forces. If the tension force continues to increase while the compression force remains the same, not much happens as the paper carries the tension load without failure, and the compression load holds steady at a value that will resist the tendency of the U-shaped ripples to spring back against it. But if the tension and compression forces are increased simultaneously and uniformly, the compression forces will begin to fold the U-shaped ripples into a series of parallel V-shaped ridges long before the tension forces increase to a level that will tear the paper as they did in the first example.

Now let's apply my three examples to the Earth during the fifth mass extinction event at the end of the Early Earth as its shape shifted from one ellipsoidal shape to another, with the Earth's crust acting as the sheet of paper. There were no areas of the Earth that corresponded to my first example, the T & T loading case, with tension in all directions. My second example, the C & C case, applies only to two very discrete areas in the vicinity of the poles of the ellipsoidal Earth. At the poles, the lines of constant latitude (those that are aligned parallel to the equator) form small diameter circles around the pole and have virtually no change in length as the Earth's shape changes. On the other hand, the lines of constant longitude (those that circle the Earth from pole to pole) all converge on the pole and shrink by about 7%, causing crustal material to pile up in a random jumble.

The pole in the middle of Pangaea, coincidentally near the place where the Himalayan Mountains would appear, and correspondingly, the pole in the middle of the primordial Pacific Ocean, is near the place where New Zealand and the South Pacific islands of Polynesia would rise up from the depths of the ocean floor. Deep ocean trenches, namely the Kermadec and Tonga Trenches, would push down deep into the Earth's mantle.

The combined two-directional tension and compression loading, the T & C case of my third example, applies to virtually all of the remaining mountain chains on Earth. In these cases, the tension loads associated with circumferential expansion parallel to the equator are more or less aligned with the mountain chain's ridges, and the compression loads associated with circumferential contraction in the direction aligned with the pole-to-pole circumference lines are perpendicular to those ridges.

Mountain Building

If the time period involved in the formation of the ridges is long enough, or the temperature of the rock is hot enough, the rock may act as a plastic material capable of bending without fracture, leaving a smooth, rounded ridgeline with few breaks along its length. If the ridge formation happens quickly or if the rock is below the temperature where it acts like a plastic, the crest of the ridge may fracture, leaving a sharp, jagged edge with fissures and multiple transverse breaks along its crest.

If the T & C loading happens along a coastline, only the lighter continental crust will form ridges, while the denser ocean floor crust will slide under the less dense continental crust. As it moves down into the hot mantle that underlies all of the Earth's crust, this subducted ocean floor crust material softens and melts, a portion of which may eventually make its way back to the surface as volcanic lava and ash that can form cone-shaped peaks.

If the T & C loading occurs mid-ocean, there are two ways the ocean floor will typically react. Unlike the coastline case, the mid-ocean case has equal density on all sides. As a result, ridges or trenches occur. If the crust fracture occurs at the low point in the undulation of the ocean floor, both sides of the fracture will dive downward, forming a trench, with the high static pressure of the magma underneath providing a clamping effect that reduces the likelihood of volcanic venting. One such trench is the Mariana Trench, located in the western Pacific Ocean south of Japan, and it has the deepest natural point in the world.

If the fracture occurs at a high point on the ocean floor, both sides of the fracture are driven upward to form an archipelago or seamount, with the high static pressure of the magma underneath providing a spreading effect that can break through and allow lava to flow upward, forming a volcanic island. The Hawaii archipelago, part of a trail of underwater mountains across the Pacific known as the Emperor Seamounts, is an excellent example of such a fracture.

With this understanding of crustal fractures and mountain building in mind, let's go back and take a more detailed look at Views A through I, shown in Figure 7. In Views D, E, and F, notice that the continents have shifted slightly from their corresponding positions in Views A, B, and C. For this movement to have occurred, stretching and compression of the Earth's crust had to be involved. In Views D, E, and F, the vertical, or pole-to-pole, circumference has been reduced by about 7%, while the horizontal, or parallel-to-the-equator, circumference has seen an increase of 0% near the poles and up to 6% at the equator. As a result of this movement, cracks have appeared, and rounded mountain-building has occurred in Pangaea. As these shape changes occurred, crustal fractures occasionally allowed lava to flow from beneath the crust upwards onto the surface. One of these lava flows, known as the Deccan Traps, occurred in India and appears to have taken 30,000 years to fully form. If it started at the very beginning of the Earth's shape shift and continued for its duration, the formation time of the Traps can be used as a good indication of the duration of the shapeshift event.

Most of the cracks (represented by dotted lines) occur where the continental shelf of present-day continents touch each other. Most of the mountains (shown as wavy solid lines) occur around the perimeter of, and close to, the pole at the center of the original Pangaean continent. These three Views, D, E, and F, are the only places I will show mountains.

Comparing View D to View A shows one crack that has developed running vertically from near the equator, upwards between North and South America, and then, with a slight jog to the left, between what is now the west coast of Northern Africa and the east coast of North America, producing significant separation. This particular fissure's proximity to the eventual location of the Yucatan Peninsula makes it the most likely source of the fractoemission spark that initiated the tube-to-Earth thunderbolt that formed the Chicxulub crater, as described at the end of Chapter 11.

Another crack runs along what is now the west coast of Southern Africa and the east coast of South America, though no significant separation has occurred. In North America, the Appalachian Mountains have appeared, probably lifting slowly during the entire shape-shifting process, running roughly parallel to what is now its eastern coast. In addition, an early

Mountain Building

uplifting of the Rocky Mountains has slowly formed close to what is now North America's western coast, and the South American continent has seen an early uplifting of the Andes Mountains along what is now its western coast. Both the Rockies and the Andes have an appearance similar to that of the Appalachians and have not yet been lifted to the lofty heights they have today. The Hawaii archipelago, part of the Emperor Seamounts, has begun to emerge from the Pacific Ocean basin in a line parallel to the Rockies and essentially equidistant from the equator.

Comparing View E to View B shows that the picture gets a little more complicated in this part of the world. The crack between Africa and South America, shown in View D, is visible on the left. A long crack runs down the center of the view, starting near what is now the Mediterranean Sea, running down through the Red Sea, taking a jog through the Gulf of Aden, and then continuing down to below the equator between the continents of Antarctica and Australia. A short crack branches off the left side of this central crack about halfway down and runs between the African and Antarctic continents to a point below the equator. A second very short crack also branches off the central crack to the right between India and Australia. In the process, Antarctica and Australia start their long southern migration to positions below the equator.

The two cracks that can be seen within the continent of Africa (near what is now its east coast) represent an early incarnation of the Great Rift Valley and are the only significant land-locked cracks shown in Figure 7. India has been pushed further into Asia, creating the initial upheaval of the Himalayas. The foundation for the Alps, the Balkan Mountains, the Caucasus, the Taurus Mountains, and the Zagros Mountains has formed along what is now the northern shore of the Mediterranean Sea and extends over to the Himalayas. Antarctica has seen the initial upheaval of the Transantarctic Mountains and the beginning of its central highlands. Australia has seen the Great Dividing Range begin to emerge.

When View F and View C are compared, the emergence of the Ural Mountains cutting across Eurasia at an angle can be seen, forming the dividing line between Europe on the upper right and Asia on the lower left. Part of the crack showing between Europe and North America is visible at the top right. The Himalayas, the chain of mountains from the Alps to the Zagros, and part of the Rockies are shown as discussed earlier in View E. The Hawaiian-Emperor seamount chain can also be seen in this view. The space between North America and Asia, which will eventually be the floor of the Arctic Ocean, is dry land in View F, offering a migratory route between the two continents that will last until 34,000 years ago. (In Chapter 19, I will describe the events that led to the end of Interim Earth, at which time the land in question sank to become the floor of the Arctic Ocean in response to the Earth's dramatic shapeshift from ellipsoidal to spherical.)

What seems to be missing in View F, at the lower edge of Eurasia, is a mountain chain parallel to the equator as can be seen on other continents in Views D and E. In fact, the ridge is there, but it does not show since it is just offshore under the ocean, with some of the volcanic activity from the seafloor subduction appearing as individual volcanoes on the mainland of Japan. What is very apparent in this view is the significant concentration of mountains around the pole at the top of the Earth, where all of its crust is in compression.

The details of mountains I have identified in Views D, E, and F show that they tend to occur in three distinct placements: first, at or near the pole due to compression from all directions; second, near the edge of Pangaea encircling the Earth due to stretching parallel to the equator, combined with compression in the pole-to-pole direction; and third, mid-continent at the Eurasian border, near the east coast of North America, and in the middle of Antarctica, again due to stretching parallel to the equator and compression in the pole-to-pole direction.

The Appalachian Mountains, shown in View D, and the Ural Mountains, shown in View F, deserve a few more comments. These two mountain chains appear today essentially unchanged from when they were formed at the end of the Early Earth, with virtually no additional mountain-building activity over the ensuing 65 million years. The result of the slow 30,000-year-long uplifting process that created them makes them appear more gentle, or, if you prefer, less rugged when compared to the present-day appearance of the other major mountain ranges I have discussed. All of the other mountains are similar in appearance to the Appalachians and Urals and would not reach their spectacular heights and jagged ridgelines until the catastrophic event that ended the Interim Earth only 34,000 years ago.

Except for the collection of mountains near the pole, all mountain chains run essentially parallel to the equator, where they are transverse to the pole-to-pole direction that would shrink as the planet changed shape. All major cracks that appear between continents run essentially vertically, which is perpendicular to the direction where the circumference would stretch as the planet changed shape.

Up until the time of the shape-shifting end of the Early Earth 65 million years ago, the continents were locked together in the shape of half an eggshell, fitted intimately to the contour of the egg-shaped molten core beneath them. They could turn in unison on an axis through the poles without changing that fit or stirring the molten core, but if they tried to turn on any other axis, they would find they were locked in place and could not move relative to the massive core. As the electrical forces in the flux tube gradually decreased non-uniformly, the length-to-diameter ratio of the pre-event ellipsoidal Earth changed from 1.6:1 to a post-event ratio of 1.4:1. With this new shape, Pangaea could no longer fit on top of the world. It may have resisted the bulging equator forces for a while, but the thin rocky crust was no match for the massive molten core as gravitational and electrical forces slowly changed its shape. As a result, Pangaea and the primordial Pacific Ocean floor cracked and buckled in response to overwhelming forces from within.

When the shape-shifting was complete, the continents were essentially and effectively separated from each other but remained in close proximity to their original positions. The individual continents, none of which completely encircled the planet, were now only marginally constrained by their fit to the new ellipsoidal shape, allowing them to move about in response to churning currents stirred up in the molten magma beneath them after the Earth's core changed its shape. These currents were much more powerful than the thermally induced convection currents that had existed since the Earth had formed, but which had been insufficient to disturb the original unfractured shape of Pangaea.

Finally, in the G, H, and I Views, we can see how 65 million years of continental drift has redistributed the individual continental pieces of a fractured Pangaea to positions that are not far removed from where they can be found today.

Part 3

A New World Begins

Five Friends

Chapter 14

The Survivors Evolve

Flowing lava, toxic gases, and doubled gravity—oh my! The transition from the Early Earth to the Interim Earth saw successive generations of Earth's inhabitants unable to survive either the gradual doubling of the Earth's surface gravity or the toxic gases from volcanoes and lava flows that periodically poisoned the air as our world changed its shape.

When that transition ended and all the dust had settled, the Interim Earth had begun. As I discuss my vision of the ensuing time period, I will trace the 64 million years or so of continuing evolution that saw only about 20% of the flora and fauna that had existed before the slow-motion catastrophe of the Early Earth's demise grow and prosper as they populated their new home, culminating in the rise of humans to become the dominant creatures on the planet. To help understand the land these survivors inherited and the ecological and geological changes that occurred during the Interim Earth, I have reproduced Views D, E, F, G, H, and I from Figure 7 back in Chapter 13 to create a new *Figure 8: Continental Drift During the Interim Earth* (shown below) for use in this chapter as I describe the shape of the Earth and the placement of the continents at the beginning and at the end of the Interim Earth.

Five Friends

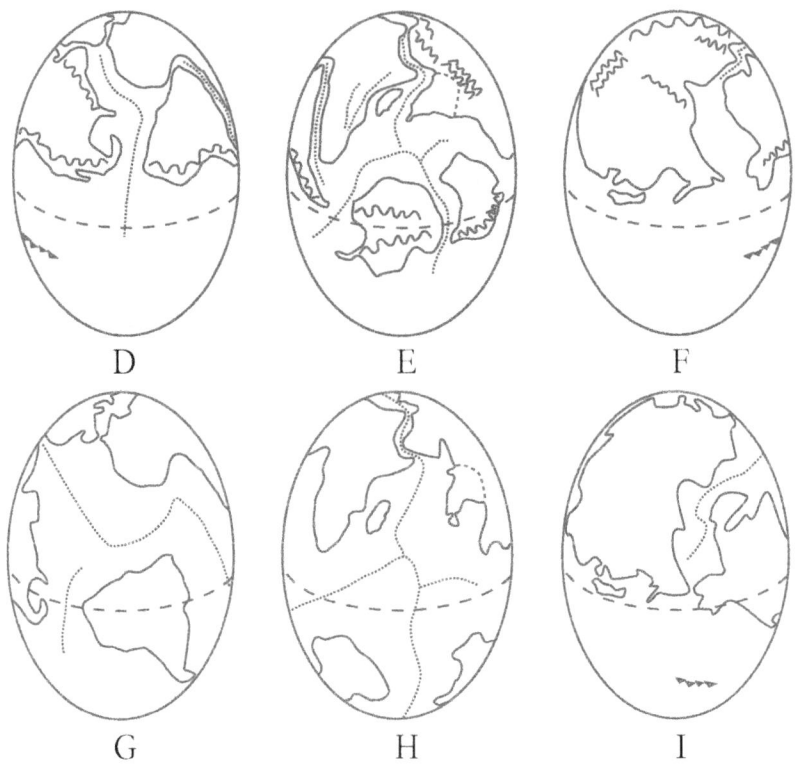

Figure 8: **Continental Drift During the Interim Earth**

The primordial supercontinent of Pangaea had been the Earth's only landmass for the entire lifetime of the Early Earth. Views D, E, and F show the Earth with a length-to-diameter ratio of approximately 1.4:1 at the beginning of the Interim Earth after its transition from the Early Earth's approximately 1.6:1 ratio. That change in shape caused Pangaea to break up into six separate continents.

Three of them—Eurasia (which includes present-day Europe, Asia, and the Indian subcontinent), North America, and Africa—though fractured from one another, all remained essentially in contact with each other, allowing free migration of land animals between them whenever the ocean level was low enough to expose the continental shelves. The remaining three continents—South America, Antarctica, and Australia—broke away completely from the original Pangaean continent and, floating on the Earth's molten mantle that was still churning from its recent shape-shifting trauma, slowly drifted away. They became completely surrounded by ocean and were virtually cut off from the other continents' migrating inhabitants. By the time the Interim Earth ended, their positions, as shown in Views G, E, and H, were not far removed from where they can be found today. For the first time in the history of the Earth, there were continental landmasses below the equator on the side of the planet that faced the Sun.

This long period of separation allowed evolutionary paths on each of the three isolated continents to develop differently from those that developed on the much larger landmass of the three interconnected continents, where competition and environmental variations led to a much more diverse population.

In the case of Australia, a small continent that has remained isolated up to the present day, marsupial and placental animals are virtually locked in a time capsule of their primitive early evolutionary state of about 65 million years ago. South America, which is more than twice the size of Australia, was isolated for 60 million years until the rise of the Central American land bridge about 3 million years ago that connected North and South America. It, too, had a primitive marsupial and placental animal population, but once the land bridge connected the two continents, more highly developed North American placental carnivores were able to invade fresh territory and become the predominant species there. It is this more diverse population that we see in South America today.

The only animal life currently living on the continent of Antarctica, which now sits literally frozen in time at the South Pole, is the flightless emperor penguin using its stable ice fields as nesting grounds. There is virtually no other animal life on that continent other than humans doing research at various outposts.

The beginning of the Interim Earth was different in many ways from that of the Early Earth. The tallest of trees no longer stood, and the less-than-23 kg (50-pound) mammals dominated the landscape. With the extinction of the sea reptiles such as the Liopleurodon, the waters of the Earth were much less threatening, allowing mammals such as whales, porpoises, and seals to colonize the sea. Bats and birds took to the air. Gone were the huge fern-like trees and the long-necked giant dinosaurs that fed on them. The flora changed from mostly lush forests to extensive grassland, savannas, and prairies in many areas. Round-ridged mountains rose up from the old undulating flatlands, reaching elevations of up to 2,100 m (7,000 feet), offering another dimension to the evolutionary niches. Volcanoes dotted the landscape.

About the only giant-sized survivor from the Early Earth was a huge snake whose fossilized remains were discovered in a South American coal mine in layers that date back to the time 10 million years after the dinosaurs' extinction. The Titanoboa cerrejonensis was about 13 m (43 feet) long and weighed about 1,100 kg (2,500 pounds), preying on giant turtles and primitive crocodiles. Its modern-day cousin, the green anaconda, only measures 7 m (23 feet) in length and weighs around 150 kg (330 pounds). The Titanoboas and the prey they lived on all benefited from their anatomical proximity to the ground and their occasional watery habitat. But even though they spent most of the time with their full body weight resting on the ground and did not need to rely on legs to continuously support their doubled body weight in the Interim Earth, their progeny eventually evolved into smaller versions of their ancestors, morphing into their modern-day equivalents.

As the jumble of fallen trees and the bodies of the no longer viable dinosaur population were slowly covered in the silt and sediment of the continuing erosion of old hills and newly formed mountains, the population of diminutive mammals not only survived but thrived. As a species, their evolutionary path radiated out into a wide array of types and sizes. Since they were exposed to only half the effective gravity that their Early Earth ancestors had experienced, they evolved with more efficient body structures that allowed them to run and jump effectively in their new environment.

The Survivors Evolve

The "bigger is better" theme that worked successfully for the dinosaurs of the Early Earth also worked well for some lines of the mammal evolutionary path during the Interim Earth, but no land mammal ever challenged the title claimed by Argentinosaurus as the largest animal to walk the Earth. The Baluchitherium, a hornless, giraffe-like rhinoceros that thrived on the Indian subcontinent between 23 and 34 million years ago, grew to a length of 8 m (26 feet), a height at the shoulders of 5.5 m (18 feet), and weighed 18,000 kg (18 long tons). It can claim the title of the largest land mammal that ever lived but would have been dwarfed by Argentinosaurus.

Figure 9: Relative Size of Animals 65 Million and 34,000 Years Ago vs. Today shown below compares the relative size of some of the animals mentioned above.

Figure 9: **Relative Size of Land Animals**
65 Million and 34,000 Years Ago vs. Today

The Argentinosaurus was the largest land animal during the time of the Early Earth. The Baluchitherium was the largest land animal during the time of the Interim Earth. The elephant is the largest land animal living today. The difference in the bulk of these three animals apparent in these images is graphic proof of the difference in the effective gravity on the surface of the Earth during its history. This memorable visual impact provides another Aha! moment. Thus,

Aha! #8:
There is visual verification of effective gravity difference.

The giraffe is the tallest and has the highest blood pressure of any land animal living today. Its image is included in this figure to graphically show that the elevation of its head above its heart, which is an indication of its blood pressure, is only about 1/3rd that of the Argentinosaurus. For the Argentinosaurus to have had a blood pressure comparable to that of a giraffe, the effective gravity during the Early Earth could not have been more than 1/3rd of today's gravity. The present-day human form is shown in this figure as a size reference.

I was tempted to include the size and weight variation of the largest flying animals as additional proof of a significant gravity increase between the Early Earth, the Interim Earth, and the present day. The flying heavyweight champion was the *Quetzalcoatius,* a 136 kg (300-pound) pterodactyloid pterosaur with a 15 m (49-foot) wingspan that became extinct at the very end of the Early Earth along with the dinosaurs. Fossil records indicate that a bird called *Palagornis sandersi,* with an estimated weight of up to 40 kg (88 pounds) and a wingspan of 6.4 m (21 feet), was the largest bird flying the skies of the Interim Earth about 20 million years ago. Today, the largest bird in the skies is the southern royal albatross, weighing up to 8.6 kg (19 pounds) with a wingspan of up to 3.5 m (11.5 feet). These reductions in flying weight closely replicate the corresponding weight reductions of the animals that walked on the Earth, lending credibility to my arguments.

However, there is another explanation for the ability of large airborne creatures to fly at the otherwise impossible weights of Quetzalcoatlus and Palagornis sandersi, and that is that the density of the atmosphere could have been significantly greater on the early Earth, allowing much greater lift from the large wings of these animals. But the land animals would have had no such advantage with the air density that would have allowed these large birds to fly. For this reason, I have limited my argument for the gravity increase over time to the fossilized land animal evidence.

As for the humans living on Earth, there were none during the Early Earth, and they only made their appearance late in the Interim Earth. A primate, similar in form to a modern tree shrew and living in the dinosaurs' world about 10 million years before the end of the Early Earth, was the earliest ancestor of modern-day human beings. By about 40 million years ago, those primates had evolved into monkeys, followed in turn by the apes, and, at about 3.75 million years ago, the hominids. The modern-day human is the only surviving species that is a direct descendant of these hominids. By 2 million years ago, these early humans had developed a brain size that was larger than that of apes, allowing them to acquire the survival skills necessary to live in a world populated by mammoths, mastodons, huge saber-toothed cats, giant armadillos, and ground sloths.

By 50,000 years ago, human intelligence had evolved significantly. Improving language skills, rapid increases in tool-making skills, and socialization with family and groups of families turned humans into a dominant species in the world. They were still hunter-gatherers living in a true Garden of Eden with fruits and edible plants in abundance. They had mastered fire and cooked the game they killed for sustenance. The superior nutritional value of the cooked food that became the centerpiece of their diet allowed them to spend less time hunting and more time in social groups with their fellow humans, developing communication skills and passing on the observations they made in their daily lives. This storytelling skill and the practice of passing down observations of significant events from generation to generation allowed their descendants, once writing skills were developed, to finally record for posterity their fantastic stories.

Five Friends

Chapter 15

An Omen and the Golden Age of Saturn

"A blinding light in the sky is an omen of disaster to come." Perhaps one day we'll find this inscribed on the wall of an ancient cave. It was a tough lesson to learn.

A key piece of evidence that I will use in my description of the events at the end of the Interim Earth is an object in the constellation Gemini that Conventional Wisdom calls a pulsar, describing it as the remnants of a supernova that occurred near our solar system about 40,300 years ago. Discovered by NASA in 1972, this object was named "Geminga." My Unconventional Wisdom, guided by the electric nature of the universe espoused by plasma cosmologists, calls this supernova a "stellar fissioning" and attributes its fluctuating radio-frequency radiation to electrical instability rather than to a rapidly rotating neutron star core as Conventional Wisdom proposes.

Regardless of the terminology, this "supernova" was an exploding star similar to the "stellar fissioning" event I described back in Chapter 8, which was critical to our own solar system's formation, but of a dramatically higher intensity. Instead of birthing a companion star, this explosion was strong enough to shatter its core, sending a shock wave and an intense pulse of light and radiation out into the universe. The light and radiation took about 300 years to reach the vicinity of our Sun. The shock wave, traveling at a much slower pace, would take several thousand years to make the same journey, sweeping a spherical region around the explosion site free of dust and gases as it plowed through the intervening space.

The light emitted in that faraway explosion was bright enough to penetrate the flux tube surrounding Earth so that it could be seen by the humans inhabiting the planet at that time. Unfortunately for those souls, they had no idea that a lethal burst of radiation accompanied that brilliant light.

All of humanity, and indeed all creatures on the Earth with eyesight, were captivated by the appearance of this strange object appearing in the continuous daylight of their sky. This was the first light that had originated outside the solar system to be seen on Earth by its inhabitants. People stopped what they were doing and stared at the light. They called their friends and family out of their shelters or cave residences and stood in awe of the vision that shone overhead. Little did they know that they were being exposed to a brief ten-second burst of intense cosmic rays that would overwhelm their bodies, create a mass extinction, and kill a significant percentage of the species, both flora and fauna, on the planet.

The light from the supernova lasted for six months as the survivors of the radiation pulse recovered from their ordeal and buried their dead, inheriting a world that was physically unchanged by the blinding light from the sky. Even after the light was no longer visible, the message it had burned indelibly into their psyche was: "A blinding light in the sky is an omen of disaster to come."

The progeny of these survivors were affected by the incident in a profound but subtle way. The genetic material in their makeup had been altered by the radiation in a beneficial way, leading to mutations that allowed their brains to grow bigger. With more thought-processing capability, their intelligence expanded to a level virtually identical to that of modern humans. They still had a lot to learn, but now they had the brainpower to develop art, music, writing, and so much more. It is these individuals to whom all of humanity on Earth can trace their DNA. They were essentially the biblical Adam and Eve.

An Omen and the Golden Age of Saturn

So far, in my description of the Unconventional Wisdom, I have always based my observations on physical evidence in the natural world. I've reviewed well-documented explanations of the Sun's measurable output that indicate it is powered by cosmic electricity. I've researched fossilized bones that date and size the extinct animals of Earth. I've studied the jigsaw puzzle pieces of continents and how they fit together. I've noticed the orientation and shape of mountain ranges that give an indication of the movement of the Earth's crust. These and many other hard facts anchor my proposals. Though I will continue to include these irrefutable bits of evidence where possible, from this point on I will also include the eyewitness observations of our earliest ancestors. Their strong tradition of storytelling and their recording of observations in symbolic art, such as indelible stone carvings, were their only means of educating their offspring and future generations in the vital survival skills they would need as they populated the world. Their myths, legends, and symbols are worthy of serious consideration, and I have not dismissed them out of hand.

It is important to note that modern-day scientists follow in our ancestors' footsteps to fulfill mankind's yearning to let the universe know we were here and what we have learned. In 1977, NASA launched two space probes, Voyagers 1 and 2, to study the outer solar system.. By 1989, their primary mission was completed, and the missions of the two spacecraft were extended since they continued to transmit useful scientific data. In 2012, Voyager 1 became the first human-made object to enter interstellar space, and in 2018, Voyager 2 followed suit.

Both spacecraft carry with them a 12-inch golden phonograph record that contains pictures and sounds of Earth, along with symbolic directions on the cover for playing the record and data detailing the location of Earth. The record is intended as a combination of a time capsule and an interstellar message to any civilization, alien or far-future human, that may recover either of the Voyagers. Hopefully, those civilizations will be advanced and "friendly." Hopefully, the information will not be considered only creative mythology but will be studied with what we would call scientific curiosity.

Five Friends

The term "The Golden Age of Saturn" has been used to describe two different eras in the history of the ancient world. The first is the one I am about to discuss, and the second, merely called "The Age of Saturn," will be discussed in Chapter 24. The first Golden Age of Saturn refers to the time period that began about 40,000 years ago and ended at the end of the Interim Earth 34,000 years ago. Knowledge of this Age has been passed down to the present day in the form of legends by those of our ancestors who survived the supernova radiation blast I have just described.

These ancients were still living on an Earth enveloped by a flux tube connecting the Sun to Proto-Saturn. They had never seen the Sun because they lived on the side of the Earth that always faced away from it. Since the gravitational level on the Earth's surface was only about two-thirds G (up from one-third G during the Early Earth), humans were able to grow to a stature larger than that of present-day humans living on a 1 G Earth. There was still constant daylight and no night. Rainbows had never been seen. The atmospheric gases trapped in the flux tube dispersed the light from the Sun to brighten the sky, thereby obscuring any view of other stars in our galaxy. The heavens looked nothing like what we see today when we step outside our caves.

Chapter 16

The Eye in the Sky

In spite of never having seen the Sun or the stars in our galaxy, these ancient cultures from many parts of the world described an object in the sky that was a constant presence. It seemed to be supplying all the light that fell on the surface of the Earth. Each separate culture revered this object as a god, giving it a unique name that would eventually be applied to the planet Saturn in their later history. These ancients referred to the Age they lived in as Golden because of the comfortable atmospheric temperatures, the abundance of fresh water, and the availability of edible fruits, vegetables, and game that were free for the taking.

The object that these ancients saw in the sky turns out to be an overlaid image of Mars, Proto-Venus, and Proto-Saturn. During the Interim Earth, planet Earth was farther from the Sun than it had been in the Early Earth. As a result, the flux tube connecting the Sun and Proto-Saturn stretched out, causing the density of the atmosphere captured therein to gradually decrease. At some point during the evolution of humans, the flux tube atmosphere thinned enough that, for the first time, these three objects—destined to become the dominant figures in their sky—became visible.

The inhabitants of what are now North Africa and the Near East, particularly Egypt along the Nile River and Iraq along the Tigris and Euphrates Rivers, lived on land near the pole of the ellipsoidal Earth, which always pointed towards Proto-Saturn. Because of their location, in their earliest history, they saw these three objects almost directly overhead in the form of a "star and crescent." Later, when the flux tube gases had thinned further, they saw it as an "eye on top of a mountain". They would reproduce these images in their rudimentary stone carvings, in their jewelry, and eventually in their sacred architecture. These images have survived to the present era in the form of the "star and crescent" symbol, which appears on the flags of many nations near the Mediterranean Sea (Algeria, Tunisia, Mauritania, Azerbaijan, Turkey, Pakistan, and Malaysia), and in the form of the "eye on top of a pyramid," which is emblazoned on the back of the United States one-dollar bill.

Before we look at these particular images in detail, it is appropriate to review the geometric relationship between the three rocky planets and Proto-Saturn to help in visualizing how their arrangement within the flux tube produced these images. Back in Chapter 9, I discussed the "gravity-only 25.3-day orbital period radius," as shown in Figure 4, depicting all five objects of the Early Solar System with their diameters and spacing shown to the same scale. During the transition to the Interim Earth, the distances between the Sun, the planets, and Proto-Saturn increased to perhaps 1.5 times the distances shown in that earlier figure.

The orbital period of the entire flux tube system increased as well, but the significance and relative position of the gravity-only orbital radius remain. It is the radius from the Sun where a planet with the same orbital period as the flux tube system would orbit freely, even if the flux tube electrical forces were not present. Planets closer to the Sun than this magic orbit would try to move with a shorter orbital period, while those farther away would try to move with a longer orbital period. This means that Earth, slightly closer to the Sun than that radius, is being ever so slightly retarded, while Mars and Proto-Venus, slightly farther from the Sun than that radius, are being ever so slightly advanced from their respective gravity-only orbits by the flux tube electrical forces, resulting in a slight distortion of the flux tube near these planets.

Figure 10, shown below, presents two views of that distortion, both of which are seen from a vantage point looking down on the plane of the counterclockwise rotation of the solar system, showing the outer portion of the flux tube, Proto-Saturn, Proto-Venus, Mars, and Earth. The Sun

The Eye in the Sky

and the inner portion of the flux tube are out of view at the bottom of the figure. The view on the right shows the diameters and spacing of the planets and Proto-Saturn to a common scale, allowing them to fit on the page. This scale makes the planet and flux tube details difficult to see but allows an appreciation of the significant distances between the four players. The view on the left shows the diameters of the planets and Proto-Saturn to a much larger scale but maintains the same distances between them as in the right-hand view to allow details of the planets, Proto-Saturn, and flux tube to be more readily seen. The relative diameter of the flux tube in both views is an approximation.

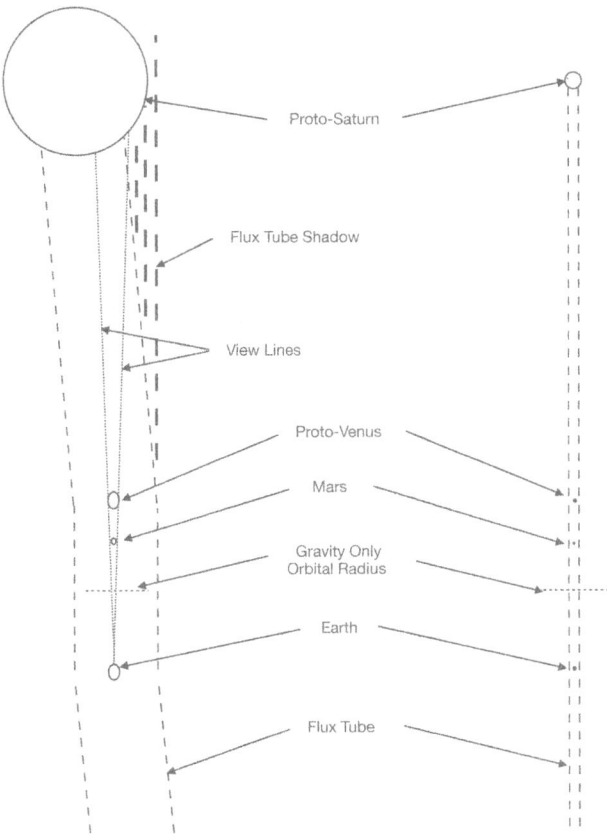

Figure 10: **Distortion of the Flux Tube by Gravity Only Orbital Effects**

In both views, the light from the Sun is shining upward from the bottom of the picture. The heavy dashed lines to the side of the flux tube in the left-hand view represent the shadow of the flux tube that sunlight casts on Proto-Saturn. The small horizontal dashed line between Mars and Earth in both views represents the location that a planet in a circular orbit around the Sun would follow under the influence of gravity alone if no flux tube electrical forces were active. The three rocky planets are shown in line with their common rotational axis, intersecting Proto-Saturn at a point positioned slightly to the right of its center. The left view clearly shows that the flux tube bends at a slight angle to the left under the influence of the forces discussed above as it maintains its alignment with the center of Proto-Saturn.

This view also shows two view lines extending from the pole of the Earth to tangent points on either side of Proto-Venus and continuing up to the surface of Proto-Saturn, showing the apparent placement of Mars and Proto-Venus against the background of Proto-Saturn as they would appear to a person standing near Earth's pole.

Suppose that person was standing at the pole facing to the left. As they lifted their head and looked straight up, they would see Mars as a slightly smaller circle centered on Proto-Venus, with the pair of them positioned directly overhead and appearing slightly above the center of Proto-Saturn. Mars, being the closest, would be the most visible, but its reddish surface would make it appear somewhat dark. Proto-Venus, at less than twice the distance away, would be somewhat less visible, but its thicker, lighter-colored local atmosphere would enhance its brightness. The portion of Proto-Saturn within the flux tube, at three times farther away, would be the least well-lit, whereas the portion in direct sunlight outside the flux tube would appear as a bright crescent shape pointing upward. The portion of Proto-Saturn that is outside the flux tube and in the full shadow of the flux tube would be virtually invisible.

The image shown in Figure 11 opposite is a representation of what that person would have seen.

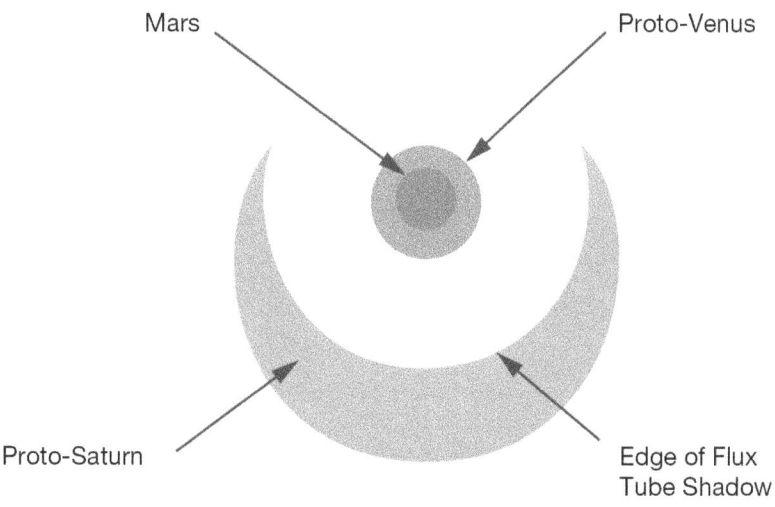

Figure 11: **The Star and Crescent Image Our Ancestors Saw**

This "star and crescent" image was created by the brightly illuminated crescent-shaped portion of Proto-Saturn that was fully illuminated by direct sunlight, in contrast to the circular dark region in the shadow of the flux tube. The combined brightness of Mars and Proto-Venus contrasted against the shadowed circular portion within the confines of the flux tube, giving the impression of a bright object within the arc of a brilliant crescent. Many ancient civilizations recorded this image, and even in the modern era, the "star and crescent" holds special significance in certain countries.

Overleaf are the flags of two nations, Pakistan and Turkey, whose territories are very close to the point on the Earth that faced Proto-Saturn during the Interim Earth, when the "star and crescent" was first observed in the sky.

Figure 12: **The Flag of Pakistan (blue background)**

Figure 13: **The Flag of Turkey (red background)**

The flag of Pakistan in Figure 12 shows a star within the extended arc of the outer diameter of a crescent, whereas the flag of Turkey in Figure 13 shows a star slightly outside the extended arc of the outer diameter of a crescent. The flag of Pakistan represents what the ancients actually saw. Four of the seven nations that use the "star and crescent" on their flags—Algeria, Tunisia, Mauritania, and Pakistan —show the star at varying positions within the extended arc of the outer diameter of the crescent.. The other three—Azerbaijan, Turkey, and Malaysia—show the star outside the arc of the crescent.. There is a subtle but very significant reason for this difference.

The Conventional Wisdom holds that the solar system has been essentially as we see it today since its creation and that any planetary alignments the ancients may have seen should be evaluated assuming that the heavens looked the same to them in their time as they do to us in ours. Unfortunately, what the ancients actually saw is not possible in today's solar system. So those who follow the Conventional Wisdom had to develop another explanation for how the "star and crescent" image could have been formed.

The only planetary objects that can display a crescent in today's solar system are Mercury, Venus, and Earth's Moon. These are the only spherical objects that can come between the Earth and the Sun, and when they do, they show a crescent shape whenever illuminated from behind by the Sun. Mercury is so small and visible for such a short time span that it can hardly be the object creating a memorable crescent, and Venus is not much better as a crescent candidate. The only answer that meets the Conventional Wisdom's requirements is that the symbol is the image of a planet or a brilliant "star" just about to be occulted by the dark portion of a waxing crescent Moon, or just emerging from occultation by the dark portion of a waning crescent Moon.

If the Moon is indeed the "crescent," then it is impossible for the star to be within the extended arc of the outer diameter of the crescent unless it is between the Earth and the Moon, allowing it to be seen in front of the solid unlit portion of the Moon. But there are no spherical objects between the Earth and the Moon. So, the Conventional Wisdom assumes that the ancients must have been wrong in their drawings of the mysterious object they saw in their sky, and they must have meant to show the star just outside the extended arc of the outer diameter of a crescent Moon, with the star probably being Venus. They further conclude that since occultations of Venus by a crescent Moon are rare and fleeting moments, the ancients must have been commemorating some historical event that occurred on one of those rare occasions. To make their flag compatible with the Conventional Wisdom, Azerbaijan, Turkey, and Malaysia drew their flag pattern with the star image just outside the extended arc of the outer diameter of the crescent image. On the other hand, Algeria, Tunisia, Mauritania, and Pakistan represented the star and crescent as it was actually seen by their ancient ancestors, and their representation is completely supported by my Unconventional Wisdom.

As time passed during the Interim Earth, the cooler temperatures associated with increased distances from the Sun caused condensation and precipitation of volatile gases within the flux tube, allowing its atmosphere to become more diffuse, just as our present-day atmosphere becomes clearer and the sky becomes bluer on cool, low-humidity days. As a result, Mars and Proto-Venus became more and more visible, and the illumination of Proto-Saturn became more uniform. The clearing skies allowed the "star and crescent" image to gradually morph into the form of a giant eye in the sky, with Mars as the pupil, Proto-Venus as the iris, and Proto-Saturn as the eyeball. The flux tube gases were still sufficiently illuminated by diffuse sunlight that the universe outside the flux tube was still invisible, just as the stars in our sky are invisible on a bright, blue-sky day. These illuminated gases caused an interesting phenomenon to become visible to inhabitants of portions of the Earth.

Take another look at the left-hand view of Figure 10 that I discussed a few pages back. In that view, the right-hand side of the flux tube is in the shade of the entire length of the flux tube between Earth and the Sun, whereas the left-hand side is fully illuminated by direct sunlight. Suppose an inhabitant of the late Interim Earth had been standing near the Earth's pole facing to the left, just as I imagined one of their earlier ancestors standing when I described the image of the "star and crescent" shown in Figure 11. As that person looked up at Proto-Saturn, the atmosphere on the brightly illuminated side of the flux tube they were facing would appear much more prominent than the heavily shaded side behind them. Remember that distances are foreshortened in this figure, so Proto-Saturn would take up a much smaller arc of the sky, making it appear perhaps twice the size of a full moon. Perspective would make the flux tube and its sunlit atmosphere appear as a giant cone with its base spread out wide at the horizon and its upper reaches tapering to the size of Proto-Saturn.

Figure 1 in Chapter 6 is the first figure I referred to in my discussion of voltages in an electrically charged solar system. It shows spiral Birkeland currents wrapped around the flux tube between the Sun and Proto-Saturn. Under some electrical current conditions, these spirals can become visible as the plasma enters the normal glow mode, much like how the plasma in a neon sign glows.. It is very likely that the improved clarity of the flux tube atmosphere allowed these spiral currents to become visible to an Earth-bound observer. The image that these atmospheric lighting effects produced was one of a giant omnipresent eye sitting on the top of a mountain. Figure 14 below is representative of the "eye in the sky" image as it appeared to our ancestors.

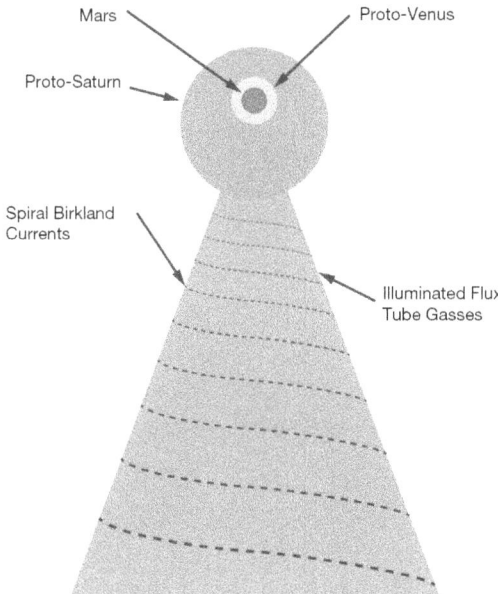

Figure 14: **The Eye in the Sky Image Our Ancestors Saw**

The ancient civilizations treated this "Eye in the Sky" as a god and recorded its image in carvings and glyphs. The spiral Birkeland currents may have appeared as a "stairway to heaven" or as the biblical "Jacob's Ladder" to our ancestors and may have been the inspiration for the Egyptian pyramids with their stone block construction.

Figure 15 below shows one of the surviving representations of this sacred image, found on the back of the dollar bill in U.S. currency.

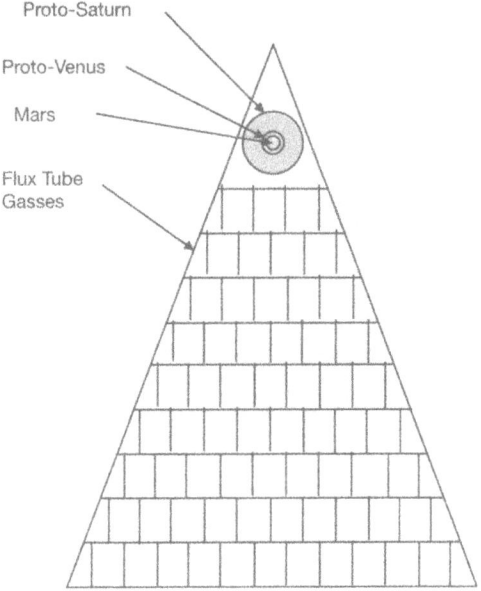

Figure 15: **The Eye on Top of a Pyramid**

Chapter 17

The Land Before Time

"Meet me here tomorrow at the same time"…would mean nothing to our earliest ancestors who lived in the latter part of the Interim Earth. They had no "night" to periodically interrupt the "day," no visible Moon, no seasons. The sky was uniformly bright, illuminated with sunlight diffused by the flux tube's atmosphere. It had been that way ever since the Earth first formed at the beginning of the Early Earth, and it would stay that way almost to the end of the Interim Earth. This was truly The Land Before Time.

Now, the Sun rises and sets on a regular basis, the Moon appears, changes phases, and disappears, and the seasons change—all marking the passage of time. Since Earth first formed, it was spinning on its axis inside the flux tube at a rate somewhat faster than the Earth currently spins. It slowed slightly when its polar moment of inertia changed as its shape shifted during the transition from the end of the Early Earth to the start of the Interim Earth. This rate of spin could have been used by the ancients as a means of measuring the passage of time, but with no visible external frame of reference, there was no way to distinguish one moment from another.

Sometime during the Interim Earth, the density of the atmosphere in the flux tube that enveloped Earth thinned enough that the sky overhead revealed to our ancestors the presence of Mars, Proto-Venus, and Proto-Saturn in the form of the images of the "Star and Crescent" or the

"Eye in the Sky," which seemed to sit atop a mountain, as I described in the preceding chapter. There were probably some among them who noticed that these images changed as they watched.

Perhaps Eve was tired and thought she would lie down to rest for a while. As she fell asleep, the last image she may have seen of the star and crescent in the sky was one with the star pointing in the direction of a particular tree near her resting spot. If, when she woke, she looked up at the sky overhead and noticed that the star no longer pointed at that tree, but had rotated about a quarter of a circle from where it had been when she went to sleep, she may have pondered, "How could that be?" It is possible that the earliest concept of time evolved as tribal members and elders discussed their similar observations of this subtle environmental nuance.

But no insight was available to the tribal elders that could have possibly prepared them for the catastrophic event awaiting them only a short time into their future. The shock wave from that faraway and long-ago stellar fissioning, which had been an omen to their ancestors nearly 7,000 years earlier, was upon them.

Part 4

Catastrophe!

Five Friends

Chapter 18

Gone in a Flash

There was nothing they could do to stop it.

34,000 years ago, a shock wave of cosmic rays made up of high-energy protons and atomic nuclei slammed into our Solar System. For the preceding 7,300 years, that expanding shell of charged particle debris, blown off by the stellar fissioning of the star Geminga, had been sweeping interstellar space free of gas and dust, leaving the volume through which it had traveled with only about 1/1,000th of the ion density in the galactic area outside the bubble it was in the process of creating.

As the vanguard ions in that shock wave ripped into the plasma of the flux tube that connected Proto-Saturn to the Sun, a brief shower of secondary particles lit up the sky of the Interim Earth in a brilliant corona glow. The inhabitants of Earth were undoubtedly startled by this ominous event as it would have brought to mind the legends of "the blinding light in the sky" their ancestors had experienced 7,000 years earlier, and the catastrophic extinction that had followed. There was indeed a new catastrophe about to commence, and there was nothing they could do to prevent or delay it. Their precious Eden-like Interim Earth and the fabled "Golden Age of Saturn" were about to end.

Within minutes, the main body of the shock wave crashed through the flux tube, distorting its magnetic and electrostatic stability to the extent that the flux tube's characteristic Birkeland currents shut down and its double-layer points disappeared. Without the restraining forces of the Birkeland currents, the flux tube's atmospheric soup of chemicals and concentrated plasma spilled out into the space around the three rocky planets and Proto-Saturn, just as the wake of the departing ion-sweeping shock wave sped away. In a flash, the flux tube was gone, never to exist again.

The voltage difference between Proto-Saturn and the Sun remained essentially unchanged after the shock wave passed, but it was still significantly less than it had been at the beginning of the Early Solar System when the flux tube first formed. As a result, the flux tube could not reestablish itself. Even though the flux tube's signature characteristic could not reform, the concentration of plasma left behind was sufficient to create a virtual plasmasphere around each planet, and a plasma tail extended from the Sun all the way to Proto-Saturn, allowing electrical forces between objects in that plasma field to interact.

The side of Proto-Saturn that faced the Sun had been the attachment point of the flux tube's Birkeland current and double-layer point force fields. With the loss of the flux tube, the electrical restraining force that had held Proto-Saturn in its much-too-close-to-the-Sun orbit suddenly decreased dramatically. The gravitational rebound, resulting from the change in these flux tube forces, combined with the impact of the shock wave on Proto-Saturn's atmosphere, caused a nova-like stellar fissioning explosion that fractured Proto-Saturn into the four large gassy planets. Today these answer to the names Jupiter, Saturn, Uranus, and Neptune. Remnants of that nova-like explosion, including a characteristic cloud of aluminum-26, can still be detected in our Solar System.

Since Proto-Saturn had been traveling close to escape velocity for its pre-explosion orbit, the new "Gang of Four" gas giants were, on average, also traveling at a velocity that would have come close to throwing them out of the Solar System if gravity had been the only restraining force acting between them and the Sun.

However, there was still sufficient plasma and a powerful (if somewhat reduced) electrical charge difference between them and the Sun to allow electrostatic forces to act, thereby constraining these four new planets so they continued to orbit at less than escape velocity.

Each planet in the Gang of Four had assumed an electrical charge similar in magnitude and sign to that of Proto-Saturn before it broke up. But since they were composed of material from different sections of their parent planet, their voltages diverged enough to cause each planet to seek slightly different orbital radii. Eventually, this tendency to seek individual orbits overcame their mutual gravitational attraction that for a while had allowed them to orbit the Sun as a group.

The interaction of their plasmaspheres produced drag forces that also contributed to their gradual pulling away from each other, leading them to enter individual, eccentric, and sometimes intersecting orbits around the Sun. These interactions, along with plasma drag forces, contributed to their gradual separation, which ultimately circularized their orbits and kept them within the Solar System. The gradual loss of charge that accompanied these interactions allowed each planet to finally settle into the gravity-only orbit in which they are found today.

When the flux tube disappeared, the three rocky planets —Earth, Mars, and Proto-Venus—were suddenly released from the well-ordered electrical force fields of the Birkeland currents and double-layer points that had kept them in synchronous orbits, overpowering their gravitational pull and giving them ellipsoidal shapes.

On Earth, the loss of these radial and axial electrical forces allowed the overwhelming force of gravity to reshape the planet. It changed from the Interim Earth's 1.4:1 ellipsoid (with an effective surface gravity of about 0.66 G) to an almost perfect sphere with a surface gravity of 1 G, all in a very short period of time, and all with mankind riding along on its surface.

This was not a repeat of the slow and relentless reshaping that the Earth had experienced during its transition from the Early Earth to the

Interim Earth, 65 million years earlier, requiring 30,000 years to complete. No, this was a sudden cataclysmic-spasm of a shape change that occurred over the course of hours after the flux tube had been blown away, with only the inertial and viscous effects of the Earth's core and shell.

The oceans would have stayed within their basins, and the plate tectonic boundaries established during that earlier event would have absorbed much of the new surface rending if the shape shift had been slow and gradual. This catastrophe was different. The oceans spilled out of their beds, causing tsunamis to circle the globe and tidal waves to roll across the land. The Earth cracked and lurched as its surface area shrank, moving continents and forming new mountain chains while adding new peaks to old ones. Mass extinctions ensued as volcanic activity dramatically increased, filling the air with smoke and ash that obscured the sky for years.

Here are some statistics for Earth's two shape shifts, the first ending the Early Earth (see Part 2, Chapter 13), and the second ending the Interim Earth (see above). The first shift caused a 6% increase in the Earth's equatorial circumference and the second an additional increase of 6%. The first shift caused a 7% decrease in Earth's pole-to-pole circumference, and the second an additional decrease of 9%. The first shift caused a 2.2% decrease in Earth's surface area, and the second an additional 1.2% decrease. It is clear that the corresponding numbers are very similar, but the rate of change made all the difference: 30,000 years for the first, less than a day for the second.

Mars, the smallest of the original three rocky planets, was the least affected structurally by its shape shift, but in the process was mortally wounded by the loss of its atmosphere and ocean. Its shape transformed from a mild ellipsoid to a sphere. This caused a previous tear that had formed in its crust at the end of the Early Solar System 65 million years earlier to double in width as volcanoes erupted and mountains were uplifted. As on Earth, the water in the Mars ocean spilled out of its basin, causing flash flood erosion of the planet's surface. Unfortunately, with the flux tube gone, the common atmosphere Mars had shared with its neighbors rapidly dissipated. The low gravity associated with Mars' small size was not sufficient to retain any significant atmosphere, allowing much of its surface water to evaporate into space.

Proto-Venus, the largest of the three, was the most dramatically altered in a one-two punch of fracture and dismemberment. Like Earth and Mars, when its elliptical shape tried to become spherical in response to the overwhelming force of gravity, Proto-Venus' crust and a portion of its upper mantle buckled and fractured. Gravitational forces held the broken crustal debris as a loose agglomeration piled up on a magma core.

Since it had always been the outermost of the three rocky planets and its speed was too high for its actual orbital distance from the Sun, the fractured Proto-Venus was thrown into a larger orbit and was strongly attracted by the mutual gravitational forces of the four newly formed gas giants. It was quickly captured by Saturn, settling into a highly elliptical orbit.

With each low-altitude pass, gravity gradients and electrical charge differences lifted loose pieces of Proto-Venus' fractured overburden into separate orbits around Saturn, where the larger pieces soon shaped into spherical bodies by their own individual gravity. Water from Proto-Venus' oceans and other condensed substances gravitated to some of these objects, forming Mercury-sized planets and over 150 spherical moon-like objects. The core of Proto-Venus became the Earth-sized planet we now call Venus.

Not to be outdone, the other three members of the Gang of Four—Jupiter, Uranus, and Neptune—swept close to Saturn, picking off individual members of its new menagerie of satellites. When the dust literally settled, here's how the Gang of Four made out: Saturn was left with numerous small moonlets, several objects the size of Earth's moon, and two other Mercury-sized remnants. One of these would indeed eventually become the planet Mercury. The other, known in legend as Phaeton, would lead a short but dramatic life before plunging into the Sun about 11,500 years ago. Its prominent ring system was not yet present. Jupiter had acquired the core of Proto-Venus in the form of the present-day Venus, four large moons, and many smaller moons. Uranus captured a large body that became Earth's Moon, along with several smaller moons. Neptune wound up with numerous small moons and one large moon orbiting in a retrograde direction.

Water from the oceans of Mars and Proto-Venus, along with condensed vapor from the defunct flux tube's atmosphere, quickly formed into spherical bodies that became moons around all four of the gas giants. Debris ranging from small boulders to moons, now known as the asteroid belt, was left behind, orbiting nearly the same distance from the Sun as Saturn when it captured and dismembered Proto-Venus.

The mechanism whereby the fractured outer layer of Proto-Venus' crust was redistributed throughout the Solar System is worthy of another "Aha!" moment:

Aha! #9:
Saturn lifted and redistributed the fractured crust of
Proto-Venus.

The fracturing itself was an inevitable consequence of the planet's final shape shift, but powerful as it was, it was not sufficient to lift all of the resulting debris off the surface far enough to escape the gravitational attraction of the remaining core. Proto-Venus' original location outside the gravity-only orbit radius meant its orbital speed was too high for that orbit once the flux tube was destroyed. As a consequence, it was thrown into a new, higher orbit that brought it close enough to Saturn for its capture and dismemberment.

During Earth's dark times following this catastrophic shape shift, human survivors could not see the magnitude of the changes that had occurred all around them—but they could sense that things had dramatically changed. Their fledgling concept of time, first conceived near the end of the Interim Earth from watching the subtle rotational movement of their "Eye in the Sky" god, helped them realize the darkness of the clouds above varied in intensity on essentially the same cycle. They could feel that the atmosphere was colder and could sense temperature cycles from warmer to cooler and back again over long periods. In some areas, they saw frozen water for the first time.

The most profound effect they could feel was the loss of electrostatic lifting forces and the resulting 52% increase in Earth's effective gravity, making even the simplest movements more strenuous. As time passed, the surviving large land animals, including the largest giant-sized humans, felt the Earth's huge gravity increase most severely and eventually succumbed to extinction. Smaller-framed humans and other animals of smaller stature, able to adapt more easily to the increased gravity, continued the evolution of the animal kingdom on Earth.

Back in Chapter 16, Figure 10, I presented the simple, well-ordered synchronous alignment of Earth, Mars, Proto-Venus, and Proto-Saturn, all within the Solar System's ubiquitous flux tube. The configuration shown in that figure is representative of the Solar System just before the flux tube was destroyed. To better understand the unseen drama playing out above the clouds and dust that surrounded Earth for several years immediately following the end of the Interim Earth, I will utilize a new graphic, Figure 16: The Solar System in a State of Flux, shown below. It depicts the entire Solar System about a month after the flux tube was destroyed as it entered into a tumultuous period of planetary billiards I have named the Transitional Solar System, which would take nearly 32,000 years to settle into the ordered stability of present day. The depicted arrangement is my best estimate based on recorded observations of our ancient ancestors, as preserved in stories, some of which I will describe in detail in Chapters 20 through 28 that follow.

Five Friends

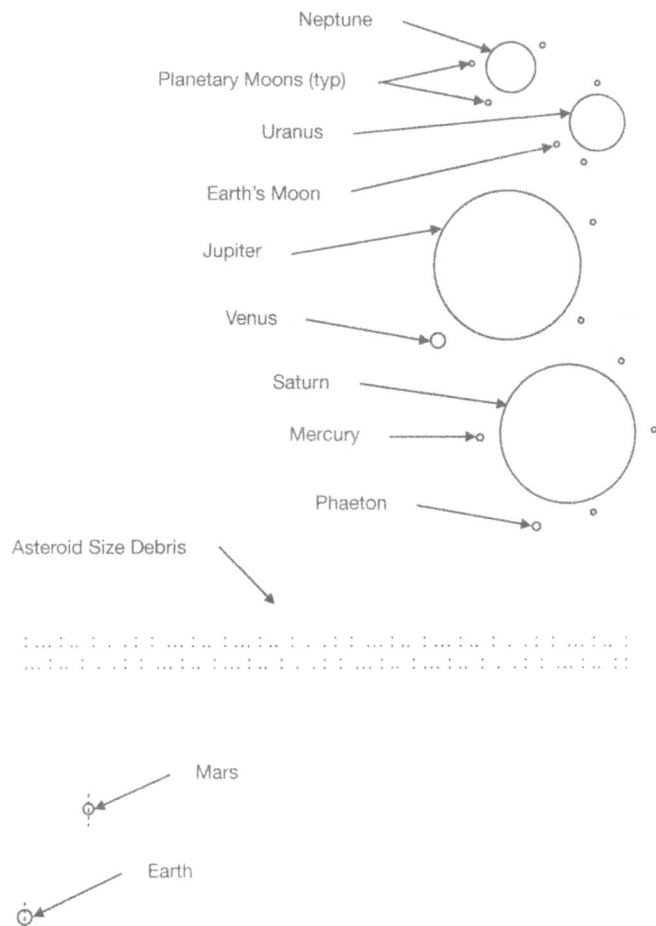

Figure 16: **The Solar System in a State of Flux**

In Figure 16 above, a chaotic scene has replaced the well-ordered planetary arrangement shown back in Figure 10. As in that previous figure, the diameters of the planets are presented approximately to scale, but the distances between objects are proportionally reduced to aid in our ability to see details. The Sun is out of view, well below the bottom of the picture.

The four remnants of Proto-Saturn—Neptune, Uranus, Jupiter, and Saturn—are shown clustered together under the effects of their mutual gravitational attraction at the upper right, in the arrangement described earlier in this chapter. Mars and Earth are shown in the lower left corner, just below the asteroid belt.

In reality, the "month after the loss of the flux tube" time frame is too short for the remnants of Proto-Venus to have achieved their orbits around the four gas giants and for the asteroid belt to have become organized. However, it is the approximate time frame necessary for the relationship between the cluster of large planets in the upper right corner and the pair of planets (Mars and Earth) in the lower left corner to have been established. This relationship is key to the following subject matter. The loss of the flux tube meant the loss of solar orbital period synchronization for these objects. As a result, Earth, even with a lower orbital velocity, has pulled ahead (to the left) of Mars due to its shorter orbital path length. For the same reason, Mars has pulled ahead of the cluster of large gas giant planets.

In my preliminary sketches for the eventual layout of Figure 16, as I planned my description of the chaotic planetary movements to come, I paid little attention to the fact that the spin axes of Earth and Mars were in the plane of the ecliptic. As I worked on my wording for the planetary interaction, I came to a screeching halt when I realized that I did not have an explanation for how the spin axes for these two planets could be inclined to their present-day 67 +/- degree positions. If I could not resolve this quandary, my entire thesis would fall apart.

Fortunately, at this critical juncture, I had another of my crucial "Aha!" moments:

Aha! #10:
Gyroscopic precession lifted Earth's and Mars' axes.

I realized that even though the flux tube was gone, the residual electrical charge distribution on all the planets would be very similar to what it had been while the flux tube was intact. In addition, plasma spheres still envelop the members of the present-day Solar System, and they were much stronger than today. The presence of this plasma allowed electrostatic forces to reach between the planets, so the strong attraction between the outward-facing pole of each planet and the average voltage of the gas giants would cause gyroscopic precession of Earth and Mars.

My only problem now was ensuring that the somewhat mystical gyroscopic precession movement would be in the right direction. To be absolutely certain I had the correct motion direction, I went to a local toy store and bought a gyroscope. I hadn't owned one since childhood and was just as fascinated by its magic as I was back then. Sure enough, the pull direction I had developed in my sketch resulted in the upward tilt of the gyroscope. *Whew!* Another close call averted.

Please use Figure 16 for reference as I describe the interactions made possible by the geometric relationship between Mars and Earth and the Gang of Four at a pivotal moment in the ordering of the Solar System. Notice that Earth and Mars are shown with a centerline running through their poles. These lines represent the rotational axes of the two planets as established while they were locked in the confines of the flux tube. These axes lie in the Sun's rotational plane (the plane of the ecliptic), and in this view, both planets are rotating such that their near-side surface is moving to the right at a rotational speed of approximately 24 hours per revolution.

Notice that no centerlines are shown for Neptune, Uranus, Jupiter, and Saturn, as their rotational axes are perpendicular to the Sun's rotational plane and are therefore being viewed end-on. If the rotational axes of Earth and Mars had remained in the Sun's rotational plane as shown in this figure, life on Earth would be dramatically different today. Both the North and South Poles would, just as they do today, see six months of total darkness followed by six months of constant daylight. However, unlike today, the Sun would be directly overhead at mid-summer, not a mere 23 degrees above the horizon. Permanent polar ice caps would not form, and the annual global climate temperature fluctuations would be so dramatic that life on Earth would be very difficult.

The planet Uranus has been oriented this way in its 84-year orbital period around the Sun for perhaps 20,000 years, demonstrating that time alone will not significantly alter the angle of a planet's rotational axis.

Back in **Chapter 6: Electricity**, where I first described the role of electricity in the formation of the Solar System, I used two illustrations (**Figure 1** and **Figure 2**) to show the structure of the electrical voltage gradient within the flux tube and its ability to distort the nascent planets contained therein into their ellipsoidal shapes. The strong electrical charge gradient that had existed between the opposite ends of both Earth and Mars during their life inside the flux tube took many years to decay after the flux tube itself was destroyed. As I have described previously in this chapter, the four gas giant planets remained close to Proto-Saturn's pre-breakup voltage. The result was a strong electrical attractive force between the outward-facing poles of both Earth and Mars and the average voltage of the Gang of Four.

In **Figure 16** on page 128, Earth and Mars are moving to the left with respect to the big four, increasing both the angle of the electrical forces between them and the distance from them with each passing day.

This planetary placement set up a one-time-only opportunity for gyroscopic precession of the rotational axes of Mars and Earth to occur. Within six months for Earth and about 12 months for Mars, the angle and magnitude of the electrical forces would no longer cooperate to tilt these two planets' rotational axes out of the Sun's rotational plane. But that was all the time they needed—the job got done. **Figure 17** (next page) depicts the Earth as the electrical forces from the four gas giant planets, acting in the plane of the Sun's rotation and at an ever-increasing angle to Earth's rotational axis, caused gyroscopic precession that tilted the axis upward from the plane of the ecliptic to an angle of approximately 67 degrees.

Five Friends

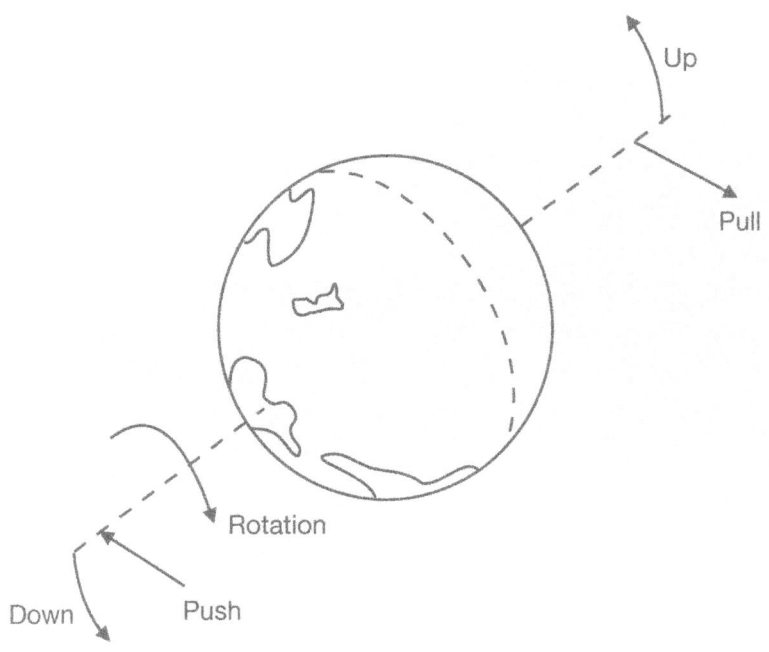

Figure 17: **Gyroscopic Precession of Earth's Rotational Axis**

In this view, we see the South Pole of the Earth, with the continent of Antarctica at the pole, the tip of South America at the bottom of the view, the eastern portion of Australia at the top, and New Zealand slightly left of center near the middle of the South Pacific Ocean. An arrow shows the Earth's "Rotation" direction. The North and South Poles at the time were charged to different voltages such that the North Pole was much more strongly attracted to the big four planets than the South Pole. The arrow marked "Pull" represents the perpendicular-to-the-axis component of the force exerted on the North Pole by the electrical attraction between Earth and the Gang of Four planets. The arrow marked "Push" represents the virtual push of the weaker pull-force exerted on the South Pole.

The moment, or torque, produced by the "Push" and "Pull" forces on the rotating planet-sized gyroscope caused the Earth to precess at a right angle to the applied torque, resulting in its North Pole moving "Up" and its South Pole moving "Down." In the short time available while these electrical forces remained effective, the tilt of Earth's axis was moved to a position close to what it is today. The forces and resulting movements I have described for Earth also applied to Mars, bringing the tilt-angle of its axis close to what exists for Mars today.

During the several years that elapsed before Earth's dark pall of volcanic smoke and dust began to lift, the four gas giants—Jupiter, Saturn, Uranus, and Neptune—moved into large erratic, occasionally crossing orbits around the Sun. The small icy and rocky asteroids started to form a ring around the Sun. The two intact planets, Mars and Earth, had moved into somewhat eccentric orbits around the Sun. Venus, Mercury, Earth's Moon, Phaeton, and all of the other moon-sized objects, with their wide variation in size and composition, were dispersed among the four gassy remnants of Proto-Saturn, each in an irregular orbit around its host.

When the dust-cloud curtain finally lifted, the human survivors of the catastrophe that had rocked planet Earth were greeted with a completely different scene, both on Earth's surface and in the sky. Aside from the doubling of the effective gravity that they had experienced immediately after Earth's final shape-shift, the most profound change was the appearance of the Sun, which they had never seen before, and the introduction of its accompanying day/night cycle, which they had never experienced, though it had been hinted at by the dark and then less-dark periods they had observed long before the clouds finally dissipated.

The stationary god-like "Eye in the Sky" that had dominated the perpetually bright sky for their entire history was gone! Now, there were at least six distinct objects that could be seen at different times during the day/night cycle, all of them moving in an arc overhead, rising at one horizon and setting at another. The largest of these objects they named Saturn, in honor of their old god that had stood steadfast overhead for their entire collective memory of the time before the cataclysm. They gave names to the other five too, calling them, in order of decreasing apparent size: Jupiter, Uranus, Sun, Mars, and Neptune.

Whenever the Sun was above the horizon, its brilliance overpowered even the larger objects in the sky and lit up the remaining dust in the atmosphere, rivaling the brightness that they had seen near the end of the Interim Earth, their Golden Age of Saturn. The landscape that daylight revealed was dramatically changed from before. Vegetation had been stripped away by the tsunami waves that had swept the land. The remains of animals killed in the cataclysm could be found jammed into caves and fissures in higher ground. Jagged peaks of newly formed mountain ranges towered over older, rounded mountain ranges and barren plains littered with tumbled rocky tsunami debris in irregular piles of rounded stones.

Once the Sun had set, the light from any of the remaining five objects still above the horizon shone brightly in the reflected light of the Sun, creating a glow in the dusty atmosphere that only the brightest stars in the galaxy could penetrate. But penetrate they did, and for the first time since the light from the stellar fissioning of Geminga 7,000 years earlier had pierced the still-intact flux tube, mankind saw objects from outside our Solar System shining in the sky.

As humans observed the movements of these new objects in the sky, their fledgling concept of time—first conceived from watching the subtle rotational movement of the Eye in the Sky formation of the Interim Earth—began to develop to a new level of sophistication. A select few among them were content to spend their lives studying and recording the steady movement of the stars and planets across the night sky, the subtle, slow, and somewhat erratic movements of the planets with respect to the star field, the movement of individual satellites of the five visible planets, and the gradual rotation of the star field itself with respect to a seemingly fixed pole star. These individuals would become prophets, able to predict future planetary actions based on past movements. They observed that the three big players—Saturn, Jupiter, and Uranus—moved in erratic orbits around the Sun, occasionally passing very close to one another, and realized that this action was setting the stage for the next catastrophe to be inflicted upon Earth.

Before we delve into the details of the Transitional Solar System described to us by ancient and modern-day storytellers, there are a few features of Venus and Mars that are observable today, which I would like to review since they add strong, evidence-based credibility to my thesis.

Five Friends

Chapter 19

Telltale Features of Mars and Venus

Battle-scarred as they are, Earth and Mars are the planets of the original three orbiting the Sun that now reveal their primordial surfaces. The largest remaining part of Proto-Venus, namely Venus itself, is showing us a surface that is only 34,000 years old, while the remains of its shed outer surface are scattered throughout the Solar System. Up until this point in my presentation, I have mentioned Mars and Venus only peripherally, spending most of my time describing happenings on Earth, about which our knowledge is much more complete. However, there are features of the surface of Mars and Venus, and the chemical composition of meteorites, that lend evidence-based data to my basic thesis.

Today's technology gives us access to computer programs that render a 3D representation of Earth based mainly on satellite imagery, and of Mars, based primarily on images recorded by the Curiosity rover that NASA sent there draped with cameras around its neck.

Mars has recently revealed features of its surface that mirror much of what I have described as having happened on Earth, thereby providing confirmation for many of my conclusions. If it happened on Earth, something similar should have happened on Mars. If it didn't, I should be able to supply a logical reason as to why not.

I spent a lot of time studying Google Earth to verify my Unconventional Wisdom explanation of our home planet's plate tectonics, primarily by observing the directions and locations of seafloor fractures and mountain ranges that I show in Figure 7. The success of that activity led me to spend some time studying Google Mars to see if I could discern how our sister planet responded to the same forces that shaped Earth. The Mars terrain features that were most helpful included the color-coded surface elevation information, the visual abundance of craters in some places and smooth surfaces in others, and signs of vigorous water erosion in some areas without the presence of a complex Earth-like river network.

There are some remarkable confirming features I would like to describe. To aid in this presentation, I have created a sketch shown below as Figure 18:

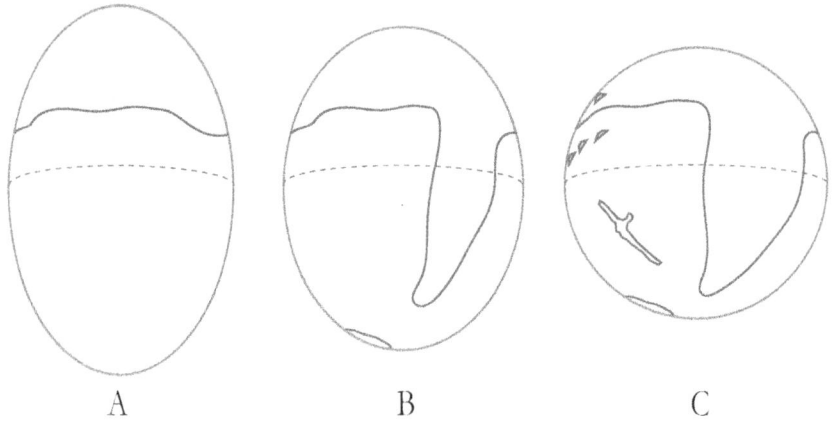

Figure 18: **The Surface of Mars 65 Million & 34,000 Years Ago vs. Today**

In Figure 18 above, View A shows my proposed geometry for Early Mars with the same ellipsoidal shape ratio (1.6:1) as I used previously for Early Earth, using the rationale that Mars was subjected to a similar ratio of axial and radial electrical forces within the flux tube as Earth, and should therefore have approximately the same shape. The dashed line represents the equator, and the North Pole is at the top. The irregular solid line above the equator represents the approximate shoreline of a primordial ocean that covered the polar region. The remaining area is dry land.

In View B, I show how the 30,000-year-long transition from Early Mars to a less elliptical Interim Mars created a tear in the crust of Mars, extending from the shoreline, across the equator, and reaching almost to the South Pole. Just as I described back in **Chapter 13** for Earth, the volume of Mars did not change during this shift, but the circumferential distances and surface area did. As with Early Earth, the ellipsoidal shape ratio for Early Mars would have changed from 1.6:1 to 1.4:1, resulting in a circumference increase of 6% in the direction parallel to the equator, most of which was compensated for by the "V"-shaped tear in the surface, and a decrease of 7% in polar circumference, which would have resulted in the uplifting of mountains across the planet.

When the flux tube was destroyed at the end of the Interim Solar System, Mars, just like Earth, transitioned to a spherical planet in a matter of hours. **View C** shows Transitional Mars in its resulting spherical shape, with the poles and equator as they would have been in the first few days of Transitional Mars. The previous tear in its crust doubled in width as the second shape-shift occurred. Volcanos erupted, and mountains were uplifted again. All of these additional features shown in **View C** can be seen on the surface of Mars today.

Now let's look at seven prominent features that can be seen today on the circular disk of Mars to help with the orientation of **View C**. The first and most important feature is the shoreline of the primordial ocean. Above the roughly horizontal portion of the shoreline lies the largest lowland region of Mars, named **Vastitas Borealis**, and within the descending "V"-shaped area are two large basins that blend into each other—the larger is named **Acidalia Planitia**, and the smaller is **Chryse Planitia**. To locate this shoreline on a Mars globe (i.e., Google Mars), draw an arc through the following points: 45°N, 72°W; 20 °S, 140°E; and 54°N, 70 °E. Then move down to 31°S, 28°W, and back up to 45°N, 72°W. Two other useful reference points are the present-day North Pole, located at 52°N, 178°E, and the South Pole, located at 52°S, 20 °W.

The second feature is the small, irregular arc to the left of center on the lower edge of the disk, representing an edge-on view of **Argyre**

Planitia, another basin that Conventional Wisdom says was caused by a huge meteor impact, but which my theory identifies as a very large sinkhole.. Two more basins that my theory also identifies as sinkholes, **Hellas Planitia** and **Isidis Planitia**, are located on the other side of the disk in **View C** and cannot be seen. More about these basins anon.

The third major feature is the very small truncated cone at the upper left (just above the shoreline), representing **Olympus Mons**, the largest volcano on Mars. Just below this feature are the fourth, fifth, and sixth features, also small truncated cones representing a chain of three volcanoes: **Arsia Mons, Pavonis Mons**, and **Ascraeus Mons**—all similar to but smaller than **Olympus Mons**, and collectively known as the **Tharsis Montes**.

The seventh feature is a long, jagged canyon located about halfway between the sinkhole feature and the equator, called *Valles Marineris*. Conventional Wisdom identifies this canyon as a fracture in the crust of Mars, but my theory suggests an entirely different origin, which will be described later in Chapter 26.

The dominant surface characteristic visible in View C is the dramatic "V"-shape in the shoreline of Mars' primordial ocean. For this shape to have materialized on Mars, the planet's shape had to have shifted from an ellipsoid with a ratio of 1.6:1 to a sphere with a 1:1 ratio. When I verified the existence and size of this "V"-shaped feature, I had a magnificent Aha! moment:

Aha! #11: Mars mirrors Earth.

The presence of a "V"-shaped cut in the edge of an otherwise uninterrupted continental shape is reminiscent of Conventional Wisdom's contention that a spherical Earth's **Pangaea** had a "V"-shaped Tethys Sea**,** as I described back in Chapter 12, Figure 6: Continental Drift— Conventional Wisdom, View J. My description there includes the words, "…shows Pangaea at 225 million years ago with the Tethys Sea straddling the equator, North America and Eurasia above the equator, and South

Telltale Features of Mars and Venus

America, Africa, India, Antarctica, and Australia below the equator." **If Figure 6, View J were rotated 90 degrees counterclockwise, its orientation would be similar to View C above.**

With all this information in hand, I began to wonder why the crust of Mars behaved differently than that of Earth during the two shape-shifting events. Whereas Mars cracked with a single split, leaving a continuous landmass with a pie-shaped piece missing, Earth broke up into six disconnected continental landmasses. What was it about Earth that made it fracture, whereas Mars simply developed a single partial tear?

I had read that scientists who have been studying Mars for a long time agree that there is no evidence for plate tectonics on Mars, and I personally saw none while poring over satellite images of Mars. So, perhaps the question should really be: Why is there no evidence for plate tectonics on Mars?

I imagined the crust of Early Mars as the shell of an egg—not too hard a stretch since they are both elliptical shapes. In addition, it turns out that the percentage of their total volume locked up in their rigid crust or shell is very similar. For Mars, that percentage is 4.4%, and for the particular egg that I measured, that percentage was 4.6%. When I calculated that same percentage for Earth, it turned out to be only 1.9%. Stated another way, Mars has 2.3 times more of its volume as solidified crust than does Earth.

The greater proportional thickness of Mars' crust prevented it from folding into the type of rounded ridges I described back in **Chapter 13**, as occurring on Earth in the slow shape-shift that ended the Early Earth. In addition, Early Mars cooled faster than Early Earth due to its smaller size and consequently higher surface-to-volume ratio. This meant that Mars' crust was cooler than that of Earth, leading to a more brittle surface on Mars that was more likely to fracture than bend.

With Earth's warmer and proportionally thinner crust, rounded ridge mountain ranges like the Appalachians and Urals formed with their ridges running parallel to the expanding equator as they absorbed the effects of the shrinking pole-to-pole circumferential shrinkage. When a tear started, it would align itself perpendicular to the equator and progress toward the pole until the stress was relieved.

On Mars, instead of bending, the cooler and proportionally thicker crust fractured, resulting in broken pieces of mountain-sized crust that bunched together to relieve compression forces as they slid relative to each other to relieve tension forces.

Imagine a square piece of paper and a thin, rigid plaster square of the same size being forced into contact with a slightly spherical surface. The paper square folds to accommodate dimensional change, whereas the plaster square fractures into small pieces that shift to accommodate dimensional changes with varying degrees of crack width between chunks. Earth—think paper; Mars—think plaster.

In addition to the crust volume percentage, there is another important difference between Earth and Mars. Early Earth was mostly ocean, while Early Mars was mostly land. This means that the primordial landmass on Earth was all on one side of the equator, whereas the Martian primordial landmass extended from one pole to beyond its equator. This subtle difference meant that as the slow shape-shift on Earth began with two forces in play—an expanding equatorial circumference and a compressing polar circumference—the single continent, **Pangaea**, would tend to be pushed away from the equator by the expansion force and pulled toward the equator by the compression. This would result in the mountain building and continent cracking described in **Chapter 13**.

When Mars was subjected to these same two forces, the portion of its landmass that extended below its equator tended to lock the entire landmass to the equator. With the polar circumference contraction restrained by the extension of the crust beyond the equator, motion toward the South Pole was prevented, creating a negative pressure in the molten magma beneath the surface. As a result, two massive sinkholes developed near the South Pole to help relieve the negative pressure between crust and mantle until a major north-south crack developed, relieving the tension in the equatorial land belt and allowing the entire landmass to slide toward the Martian South Pole.

Once freed from its equatorial belt, the landmass began to fracture to accommodate radius of curvature changes, while the single "V"-shaped crack accommodated circumferential expansion. The chunky structure of the fractured landscape allowed lava to seep up and solidify to form a single rigid shell during the 64 million years of Interim Mars. When Interim Mars ended with the destruction of the flux tube, the final sudden shape-shift on Mars resulted in the doubling of the width of its "V"-shaped crack and perhaps the addition of another sinkhole, but still no plate tectonics.

This description explains why multiple continents never formed on Mars to perform plate tectonic movements, creating the basis of another Aha! moment:

Aha! #12:
Sinkholes, but no continental drift on Mars.

Before we leave the Mars discussion, I would like to describe two other observations I made while studying Google Mars. The first is the presence of what appear to be several moderate to massive water erosion gullies that look similar to flash-flood runoff erosion here on Earth. These gullies can be seen along the sides of the "V"-shaped crack and in a few other places along the polar region shoreline. Water sloshing out of, and running back into, the ocean basin during the final sudden shape-shift of Mars would cause such erosion. The floors of some of these gullies have impact craters, indicating that the flooding occurred before the meteor showers that created the craters.

The second observation is that the area within the shoreline of my proposed ocean on Mars appears remarkably free of impact craters. This is what you would expect if, at the time of any meteor showers, there had been water within that shoreline to absorb the impact of crashing meteorites. There is also a dearth of impact craters and water erosion gullies around the four large volcanoes shown in **View C** in the area between **Olympus Mons** and the original South Pole. With respect to the present-day poles, this area lies southeast of the volcanoes in exactly the direction where a northwesterly wind would blow volcanic ash from those volcanoes. This ash would tend to cover and obscure any existing gullies and impact craters in the downwind area, but only if the eruptions occurred after the impact craters were formed. An examination of **the Google Mars** maps shows several partially covered craters at the edge of the crater-free area, confirming this explanation.

These two observations tell me that the water erosion gullies and ash-producing volcanoes formed during Mars' shape shift, while the craters on the surface of Mars are the results of a short-lived intense meteor shower that started as the asteroid belt was forming shortly after Venus was stripped of its fractured surface while in orbit around Saturn, as described earlier in this chapter.

If Mars had not been protected by a flux tube, it could have been struck by meteors at any time. However, without the flux tube to maintain suitable atmospheric pressure, liquid water could not have formed an ocean on Mars, and the area within my proposed shoreline would be just as pockmarked by meteor craters as the rest of the planet. If the large volcanoes that produced the ash currently covering the area to their southeast were not active after the flux tube was blown away and after the intense meteor showers ceased, the area described would also be pockmarked with craters.

Here is my takeaway: Most of the meteor craters on Mars were made over a short period of time, probably measured in tens of years around 34,000 years ago. This is perfectly compatible with my thesis and represents another of the marvelous Aha! moments that have occurred during the writing of this book. Thus,

Aha! #13:
There were short-term heavy meteor showers
on Mars around 34,000 years ago.

I will conclude this chapter with observations about present-day Venus and meteorites. Venus today has a diameter that is 5% smaller than Earth's. One of the first calculations I made as I developed my thesis was to determine the volume of Proto-Venus. I started with the assumption that Proto-Venus had a volume equal to that of present-day Venus plus the volume of all the rocky material in the Solar System that is contained in present-day Mercury, Pluto, Charon, the Earth's Moon, the asteroids, the moons of the gas giant planets, and the no-longer-existent Phaeton. The result was a Proto-Venus with a diameter about 10% larger than Earth's.

Even though the surface of Proto-Venus is no longer available for direct observation, two prominent features observable on the surface of present-day Venus support my thesis: its abnormally high surface temperature and its lack of impact craters.

I concluded that if all the material included in my calculations had its origins in the lithosphere and asthenosphere (the crust and upper layer of a planet's mantle) of **Proto-Venus**, it would have created an additional thickness of about 677 km on top of the surface of a core the size of present-day Venus. The crust part of this thickness would have been rigid, and although the asthenosphere part would have acted in a ductile fashion for slow strain rates, it too would have become rigid if subjected to high strain rates. This means that all of the material I have described would have fractured if exposed to Proto-Venus' rapid shape change as my thesis proposes, making it easily lifted from the Venus-sized core during its disassembly while in orbit around Saturn.

On Earth, the base of the asthenosphere has a temperature of about 1,900 K and lies at a depth of about 700 km. It cannot be a coincidence that the lower asthenosphere boundary and the depth of shattered material from the surface **of** Proto-Venus are virtually identical. The temperature of the newly exposed surface would make that surface fluid to the extent that any impact craters would be short-lived. In addition, the present-day mean surface temperature of Venus at 735 K is not an unreasonable cool-down temperature to expect for a 1,900 K Venus-sized planetary surface exposed to deep-space cooling for 34,000 years.

Yes, we have another Aha! moment:

Aha! #14,
Present-day Venus' surface temperature and lack of impact craters are compatible with Proto-Venus having shed its original surface at the end of the Interim Earth.

Meteorites reveal another interesting bit of corroborative information worthy of another Aha! moment. When Proto-Venus was stripped of its loose surface debris while in orbit around Saturn, some of those rocks included material from at least 70 km below the surface, where temperatures and pressures were high enough to produce iridium, an element commonly found in meteorites at concentrations 500 to 5,000 times that of crustal rock. This provides strong evidence that meteorites came from Proto-Venus debris. Thus,

Aha! #15:
Iridium in meteorites came from the shed shell of Proto-Venus.

Chapter 20

The Storytellers

Curious, I began to explore some "Once upon a time" stories. Could ancient ancestors shed light on any of the significant global events I was researching? What I discovered is fascinating.

We do not need to rely on physical evidence alone to understand the story of the aftermath of the catastrophic event that ended the Interim Earth. Unlike the end of the Early Earth, eyewitness accounts from the survivors of the event abound in ancient texts and symbols, telling the present world what they saw in their sky and on the face of their Earth. Their stories supply many details of the dramatic events that punctuated the more than 31,000 years of stabilization that eventually transformed a chaotic solar system into the stable and orderly system that exists today.

Our ancestors passed their observations down to their progeny through whatever means of communication they possessed. The record of the earliest years relied primarily on legends and storytelling. As their writing skills developed, the stories of the later years were recorded as carved glyphs and eventually as written manuscripts. The observations of a particular group of humans do not always agree with those of other groups witnessing the same event, due to different interpretations of the incident and different vantage points from which the observations were made, due to their geographical location differences.

The ancient and prehistoric world is a much-studied subject. Authors throughout the years have collected and interpreted the stories of the ancients as they prepared scholarly writings for the modern world to use as glimpses back into these early historic periods, which are shrouded in the mist of millenniums long past. Some of these authors have combined the physical evidence preserved in the debris left behind as catastrophic events swept the Earth with the stories of the people alive at the time, presenting a more comprehensive picture. I have used the following well-referenced unpublished manuscript and two well-documented and extensively referenced books to guide my descriptions and lend credibility to my presentation of the traumatic events that occurred during these transitional years.

In the Beginning

The manuscript for ***In the Beginning*** by Immanuel Velikovsky has never been published. The backstory as to why is unique in American publishing history. Let's begin with the book's author: He was born in Russia (Belarus) in 1895, earned a medical degree from the University of Moscow in 1921, practiced in Palestine from 1924 to 1939 as a general practitioner and psychiatrist, and moved to New York City in 1939 just as WWII was starting. He spent time at the Columbia University Library virtually every day for ten years, immersing himself in the study of ancient history.

In the late 1940s, he began writing material for a book later published as *Ages in Chaos*, which attempted to rearrange and coordinate the chronology of ancient Egypt and ancient Israel. The premise of the book is that the histories of these two countries were out of sync by five centuries, give or take. He found certain curiously similar epic events recorded in both histories, assumed they were in fact the same event, and used the relative position of the event within the individual histories to synchronize the two separate timelines. Velikovsky became curious as to whether any of these same epic events would be found in the recorded history of other ancient cultures.

The Storytellers

As he researched the mythology of civilizations from Mayan, Incan, Chinese, Indian, and North American aboriginal tribes, as well as from other far-reaching corners of the Earth, he found that all these early cultures recorded very similar celestial events. These events would have been visible worldwide at the time, though perhaps in different areas of the sky. His logical conclusion was that all these stories must have been recordings of the same events.

Velikovsky decided to put *Ages in Chaos* on hold while he wrote a book about these commonly witnessed events. He called this new book *Worlds in Collision*, and divided it into two parts: Venus and Mars. Part 1 included stories from the Bible's book of Exodus about a time 3,500 years ago when Venus was a young planet with many comet-like features in a solar system that looked nothing like what we see today. These stories describe how Venus made several close encounters with Earth, causing, among other things, the parting of the Red Sea, allowing Moses to escape from the Egyptian army, and making the Sun appear to stand still during the Battle of Jericho. Part 2 included tales from a time 2,800 to 2,700 years ago when Mars made close encounters with both Venus and Earth, as told in Homer's epic tale, the *Iliad*.

"In the Beginning" was a section originally included by Velikovsky that described events that preceded those attributed to Venus and Mars. He chose the name in deference to the opening words of the Book of Genesis, since the section described cosmic events depicted in the first book of the Hebrew Bible. However, at the suggestion of a reader at the publisher Macmillan Company in New York, who wanted to simplify the story, the material from Genesis—the "In the Beginning" part—was deleted.

Worlds in Collision was so controversial that a threatened boycott of Macmillan's extensive textbook line by powerful academics caused Macmillan to transfer the publication to Doubleday. The book became a bestseller even before that transfer took place. In 1952, after a move to Princeton, New Jersey, Velikovsky published the first part of his *Ages in Chaos* book. Then in 1955, responding to unrelenting criticism of his extensive use of myth and legend in *Worlds in Collision*, he published *Earth in Upheaval*, a work that described the same time period using extensive geological evidence to support his theories.

In late 1953 through early 1955, his friend Albert Einstein read Velikovsky's work and had many discussions with him on various subjects. Though they started with widely differing views, they grew closer to agreement during that time. It should be noted that Velikovsky was a proponent of a major role for electrical forces in the movement of stars and planets, whereas Einstein's classic description of the universe proposed that mass warped the space-time continuum, causing planets to orbit stars by gravitational forces alone. These differences must have led to several very interesting debates. Einstein had a copy of *Worlds in Collision* at his bedside when he died, just nine days after his last meeting with Velikovsky.

In the 1950s and early 1960s, Velikovsky was ostracized by most colleges and universities, but eventually became a popular lecturer. Reading Velikovsky's published works in light of the scientific findings that the intervening years produced added credibility to the claims made in these books. For instance, the data sent back to Earth by Mariner II in 1962, indicating that the surface of Venus is very hot (800 degrees F), validated the youth of the planet, one of the key revelations first published in *Worlds in Collision*.

Velikovsky spent most of his remaining years rebutting the criticism of his detractors. He died in 1979, never having published *In the Beginning*. Recognizing the importance of his life's work, his descendants have collected and catalogued his writings, making them available to all through a website titled "varchive.org."

The Cycle of Cosmic Catastrophes

The event that caused the end of the Interim Earth is most thoroughly described in a book titled *The Cycle of Cosmic Catastrophes*, by Richard Firestone, Allen West, and Simon Warwick-Smith, a well-documented dissertation on flood, fire, and famine in the history of civilization. The authors present a thorough review of Paleo-Indian/Clovis culture in North America with a detailed investigation of iron and carbon/glass particles that populate a distinct stratum of sedimentation called the Clovis Layer, appearing just below a black mat layer dating to approximately 13,000 years ago. No signs of Clovis culture exist above these layers for about 1,000 years. Throughout their discussion, they espouse the Conventional Wisdom with regard to cosmology and geology, including extensive references to Ice Age Earth surface features, which I will discuss in reference to the book *Cataclysm!* below.

They conclude that a supernova, the remnants of which are a pulsar named Geminga, exploded as close as 100 light-years from Earth about 41,000 years ago, producing a burst of radiation that caused extensive extinctions in parts of the world. They propose that the initial radiation burst was followed by a shockwave of slower-traveling ions and small particles that reached the Earth about 34,000 years ago, causing additional extinctions. They continue by proposing that a second weaker debris wave, one barely strong enough to dislodge a few comets from their orbits, struck the solar system about 16,000 years ago, and that those comets took an additional 3,000 years to spiral inward to Earth's orbit, where they made multiple impacts with the Earth about 13,000 years ago.

The authors' most compelling evidence of the location and timing of that supernova is an immense galactic chimney containing our Sun and cutting through the plane of the Milky Way Galaxy, swept clean of gas and dust by the supernova shockwave that cut through our solar system 34,000 years ago. This "Local Bubble," as it was named by astronomers, contains only about 1/1,000th of the ion density of the galactic area outside the bubble, and its diameter of about 350 light-years is completely compatible with their timeline, and with mine.

Artifacts such as magnetic particles found in the tusks of a mammoth, affectionately called "Big Ed" by the authors, are dated to the time period of the shockwave that hit Earth 34,000 years ago. The authors suppose that some of the objects therein were large enough to penetrate Earth's atmosphere and explode in the air or upon contact with the ground, sending out microscopic shrapnel that decimated herds of animals, including Big Ed. Other artifacts attributed to the second weaker debris wave, such as magnetic grains, magnetic spherules, and carbon glass, are found within the Clovis Layer that dates to 13,000 years ago.

There is a problem with the measured concentration of artifacts that they attribute to these two scenarios: The documented concentrations are much higher than would be compatible with a uniformly expanding sphere of debris from the Geminga supernova. For example, considering just the **Clovis Layer** material, assuming a star equal in volume to our Sun and located 100 light-years away was completely destroyed, the concentration of its debris would be so diffuse when it reached Earth's surface that it would make up only 0.00000002% of the 5 cm thickness of the **Clovis Layer,** rather than the 0.3% concentration of the samples collected by the authors.

The authors acknowledge this problem by describing the tendency of supernova debris to clump, thereby increasing its local concentration. They also attribute part of the debris to the previously mentioned comets that were jarred from their orbits. The scenario I have been pursuing provides an alternate mechanism for producing the concentrations the authors actually discovered.

I have already discussed that when the flux tube was destroyed 34,000 years ago, Proto-Venus and Proto-Saturn were both shattered. The rocky debris from Proto-Venus spread throughout the solar system, making it a more likely source of the large-sized rocky and icy objects that slammed into Earth in that time frame than the Geminga explosion itself. Similarly, the supernova-like explosion of Proto-Saturn that occurred at the same time, and, as I will soon discuss, a second stellar fissioning of Saturn itself 13,000 or so years ago, provides a much more credible source from within the solar system capable of producing these artifacts with the concentration and timing the authors actually discovered. Both of these explosions, occurring only a few light-minutes away from Earth rather than more than 100 light-years away, add credibility to my hypothesis.

These authors make reference to specific legends of the following tribes to make their case for a catastrophe of flood, fire, and famine around 13,000 years ago: the Lakota, Ojibwa, Hopi, Matamuskeet, Iroquois, Pawnee, Navajo, and Kato, all from North America; and the Atayal from Taiwan, the Arawak from the Caribbean, the Aztec and Toltec from Mexico, and the Inca of Peru. A review of each story they present shows a persistent problem with interpreting legends: They are often told allegorically and lack specificity regarding when the events occurred. . All ancient cultures have recorded catastrophes so world-altering that resetting calendars and redefining the positions of heavenly objects was required, often leading to the assignment of different names to the periods of time punctuated by these events. Deciding which age (and therefore which catastrophe) is being described in a given legend is open to differing opinions.

The authors reference ice core data from both Greenland and Antarctic ice sheet drilling sites several times throughout their book to validate the timing of the events they describe. A review of the source for some of this data shows that counting visual layers, delineated by inclusions of volcanic ash and the isotopic composition of the ice in the core itself to determine the age of an ice field, can produce results that range from 3. 6 years per meter to 26 years per meter for simple optical counting, and up to 240 years per meter using more sophisticated layer scanning techniques.

In the Unconventional Wisdom timeline, the Earth had its South Pole shaded by a flux tube and pointed at the Sun until 34,000 years ago, when the flux tube was destroyed and the Earth's axis tilted out of the plane of the Sun's rotation. In my scenario, the accumulation of ice at the poles could not have begun until that time. Since the Antarctic ice sheet is about 3,000 meters deep at its thickest location, the average annual accumulation rate, according to my thesis, should have been around 10 annual layers per meter. This rate of 10 y/m falls well within the observed range for the visual layer counting method cited above. As I will show in the following sections of this chapter, several catastrophes occurred in the transition period between the end of the Interim Earth and the beginning of the Present Era, which were capable of producing volcanic activity and active isotopic depositions multiple times per year over many years. These global events could easily have caused multiple depositions of visible ice core layers in a given year, leading an observer to interpret these data as annual layer boundaries, yielding ice core age projections that are 10 to even 20 times higher than my maximum proposed age of 34,000 years. This scenario appears to be the case with the data the authors cite.

Cataclysm!

An intriguing book by D. S. Allan and J. B. Delair titled *Cataclysm!* presents "compelling evidence of a cosmic catastrophe in 9500 B. C." that tends to confirm much of the material presented in Velikovsky's *Earth in Upheaval*. Allan and Delair thoroughly explore geological observations and descriptions of physical evidence, such as bones and vegetation debris crammed into caves and rock fissures, that could only be explained as the result of floodwaters.

They describe animal and plant remains in Alaska and Siberia piled in chaotic disarray, including an intact mammoth with edible flesh on its bones. These remains were found frozen much farther north than the present-day latitude where the plant remains and undigested food in the mammoth's stomach would normally grow, indicating that the entire scene was flash-frozen after being struck by a catastrophe that included a pole shift for the planet's rotational axis.

Though the authors generally stay within the Conventional Wisdom version of a very old, well-ordered Solar System leading up to the cataclysm they describe, they challenge the views held by uniformitarian Ice Age advocates. These advocates attribute the collection of linear piles of rocks and erratic boulders found worldwide to an elaborate series of terminal moraines produced by advancing and retreating glacial ice flows. Instead, Allan and Delair propose that these features can be more logically explained as the result of massive water flows from repeated global flooding. In the process, they reveal the inconvenient truth that the Ice Age as such never really happened and offer a revised timeline for the period that Conventional Wisdom allocates to the Ice Age. Among their arguments is the interesting observation that there are no rocky glacier-like moraines at the edge of the Antarctic ice sheet, which happens to represent the closest thing to Ice Age conditions that exist on Earth today.

It is appropriate at this juncture to take a closer look at the rocky geological remnants attributed to the Ice Age to see how they compare to the three distinct ways that mountainside glaciers on Earth today transport rocks. The first method occurs as the glacier plows downhill under the influence of gravity, pushing any rocks it encounters along the way, much like a bulldozer. The moving rocks tend to pile on top of each other and move *en masse* without constantly tumbling, allowing the rocks on the bottom of the pile to slide along the underlying terrain, thereby scouring the bedrock. The second method, called **plucking**, occurs when a boulder breaks loose from the ground and becomes trapped in the sliding ice. Again, the rock does not tumble but slides along the bedrock, wearing a flat surface on its underside. The third method happens as the sides of the glacier grind against the confining walls of a mountain valley, producing an undercut. Overhanging rocks break free, fall onto, and ride along the glacier's surface, forming features called *medial moraines* as the ice moves along. Again, no tumbling is involved. Since all these rocks are freshly broken from the bedrock of a valley or mountainside by a glacier, they tend to have sharp corners and edges.

As the ice melts and retreats during warmer periods, the pile of rocks at the front and edges of today's glaciers is left behind as terminal and lateral moraines. The rocks transported by plucking, and those in the medial moraines (which sometimes include very large boulders called *erratics*) are scattered in the space between the terminal moraines and the retreating ice face. After melting, glacial ice water carves out caves in the bottom of the ice as it runs downhill, reemerging at the glacier's retreating face. Small rocks and sediment swept along by these streams sometimes form long, thin piles of debris within these ice caves called *drumlins* when exposed by the retreating ice. Large chunks of ice that break from the face of the glacier often remain unmelted long enough to be surrounded by silt and sediment. The holes they leave behind when they finally melt are called *kettles*, which sometimes fill with meltwater to form *kettle ponds*.

Scientists who are proponents of the Ice Age propose that erratic boulders, drumlins, kettle ponds, and lateral and terminal moraines were created by massive glacier-like ice flows but are at a loss to explain the rounded corners of the rocks involved. Rocks transported by rapidly moving water tumble against each other, regardless of their depth in the moving mass, resulting in a pile of well-rounded rocks being deposited as the water flow rate subsides. Unfortunately, the melting of a glacier happens very slowly, resulting in small rivulets building into streams and eventually rivers. Rocks that are washed into these localized high-water flow channels may indeed have their sharp edges dulled by tumbling action. Occasionally, ice dams form, collect water, and then burst, sending a torrent of water down a valley and tumbling local rocks. However, these two mechanisms cannot account for virtually every rock in a large geographical area being completely rounded.

Any trip through the countryside of New England reveals piles of rounded rocks plucked from farmers' fields and built into walls along field edges by early settlers as they tried to eke out a living from the rocky soil of their new homeland. These rocks are not the residue of an Ice Age; they are instead the residue of mega-floods that have occurred several times since the end of the Interim Earth some 34,000 years ago.

Ice Age proponents also propose that evidence of massive ice sheets is preserved in the sloping ancient shorelines of lakes formed by glacial runoff. They attribute these features to the weight of the ice sheets depressing the Earth's crust, then rebounding after the ice sheet melts—a process called isostatic rebound **or** post-glacial rebound.

Lake Agassiz is one such lake that at one time covered much of Manitoba, northwestern Ontario, northern Minnesota, eastern North

Dakota, and Saskatchewan. My thesis attributes these sloping shorelines to the crustal curvature change that occurred as the Earth shifted from a 1.4:1 ellipsoid to a sphere at the end of the Interim Earth, a deduction that deserves another Aha! moment. Thus:

Aha! #16:
Crustal curvature change is observable in ancient dry lakebed shorelines.

The geological timeline I have been developing for my Unconventional Wisdom describes a world without any surface ice until after the end of the Interim Earth, when the flux tube was destroyed, and the Earth's axis shifted out of the Sun's rotational plane, allowing a winter season wherein water could freeze. Virtually every loose rock on the planet was tumbled by the tsunami floods I described as part of the catastrophe when the Earth shifted from its Interim Earth ellipsoidal shape to its present-day spherical shape.

As a result, except at the highest elevations, most of the mountain valley glacial moraines seen today are composed of rocks left in those valleys after being rounded by those tsunami floods or one of the great floods that followed during various planetary interactions I will describe in later chapters.

The key event described in *Cataclysm!* is the biblical Noah's flood, noted in Chapter 7 of Genesis: "The waters rose and covered the mountains…" Allan and Delair attribute the motivational force for this flood to a rogue planet called Phaeton, which they propose was created as major debris from the explosion of a supernova called Vela, located 45 light-years or so from our Sun about 12,000 years ago, and took as long as 1,000 years to reach our solar system at about 1% the speed of light. Using a complicated Akkadian epic from the ancient Sumerian region of Mesopotamia (modern-day Iran) as a reference, they describe how Phaeton was pulled into the solar system by Neptune, deflected by Uranus, and essentially captured by Saturn. It remained near these three giant planets for a long time while interacting with fractured remnants of a rocky planet (perhaps the remnants of Proto-Venus, which in our scenario included Phaeton) before heading further inward past Mars toward Earth. It passed very close to the Earth's North Pole before continuing inward and finally crashing into the Sun. As it approached Earth, debris from the outer solar system dragged along by Phaeton crashed into Earth, cratering its surface. At its closest approach, its gravitational field lifted ocean water toward the North Pole, causing the biblical Noah's flood.

In today's nighttime sky, Mars, Jupiter, and Saturn appear as unimpressive tiny dots of light that can be spotted by the unaided eye if you know where to look. However, to find Uranus and Neptune or to see the moons of Jupiter and the rings of Saturn, a telescope or strong binoculars is required. One intriguing aspect of the foregoing description of Phaeton's path through the solar system is that the ancient Mesopotamians must have been able to clearly observe the detailed motion of the large outer planets and smaller Mars-sized planets to describe their story in the extensive detail that Allan and Delair report. This suggests that all these outer planets were much closer to Earth 11,500 years ago than they are today. It is also intriguing when you realize that this proximity of the four giant planets to Earth, almost 22,500 years after their creation as Proto-Saturn's offspring, shows the remarkable durability of the electrical charge difference between them and the Sun. The resulting electrical forces kept these planets in orbit much closer to the Sun than they are today, prior to the dissipation of their electrical charge.

An analysis of some characteristics that Allan and Delair attribute to Phaeton poses a problem with their assertion that Phaeton was an intruder from outer space. First, the probability that a planet-sized chunk of debris from a supernova 45 light-years away would intersect our solar system at all is extraordinarily small. Second, even at their slowest proposed speed of 3,000 km/sec (0.1% of light speed), Phaeton would have been traveling at 4.12 times the 728 km/sec escape velocity for an orbit close to Neptune, and many times more than the solar escape velocity from the

Sun in the outer reaches of the solar system. This makes capturing such an object virtually impossible.

Except for the source of the planet Phaeton, virtually all the elements of the story told in *Cataclysm!* are consistent with my description of the chaotic solar system after the end of the Interim Earth. My earlier discussion of *The Cycle of Cosmic Catastrophes* is similarly useful for understanding that critical period. However, since both references focus primarily on the Noah's flood event between 11,500 and 13,000 years ago, they will not be very helpful in understanding the early stages of the complicated sorting-out process that defined the 31,300 years of stabilization, which eventually transformed a chaotic solar system into today's stable and orderly system. For clarification of that critical time period, I will rely heavily on Velikovsky's "In the Beginning," using Allan and Delair's thoroughly researched date of 11,500 years ago for the time of the deluge.

Five Friends

Chapter 21

Early Stories and the Ages of Earth

What person in history would I most want to meet and have lunch with? At this point in my life, I'd probably say Immanuel Velikovsky. As the kindred spirits I imagine we would be, the running theme of our conversation would certainly be "Unconventional Wisdom."

Starting with this chapter, Chapter 21, and continuing through Chapter 28, I will present selected verbatim quotes from Velikovsky's unpublished work, "In the Beginning." These 124 quotes have been grouped based on the planetary age they address, with the chapter of origin from "In the Beginning" noted at the end of each quotation. I will show how these insightful writings from the 1940s align with, and add enlightening details to, the Unconventional Wisdom I've been describing throughout this book. They do not, however, fit the Conventional Wisdom at all, making them far too controversial for inclusion in Velikovsky's *Worlds in Collision* when it was first published in 1950.

In this chapter, I will first present several quotes that refer to the period just before and immediately after the catastrophe that ended the Interim Earth, and then review quotes from Velikovsky that describe the foundations of the planetary age naming system.

At the end of the Interim Earth, humankind was experiencing a true golden age. As a reminder, I'll briefly review the conditions that prevailed at that time. The "Eye in the Sky," dominated by Proto-Saturn and supplemented by Mars and Proto-Venus, was humanity's primary deity.

An enveloping flux tube shielded Earth and its inhabitants from the harsh cosmic ray environment of interstellar space. Though concentrated by gravity at Earth's surface, the atmosphere was shared by Earth's neighbors, Mars and Proto-Venus. Magnetically confined fusion in the Sun's intensely hot photosphere created atomic elements in various proportions, which were pumped into the Sun's end of the flux tube, distributing them as seed material for synthesizing many chemical compounds that enriched the atmosphere.

On Earth, there was perpetual light and warmth—no night, no seasons. The end of our egg-shaped planet, the end facing the Sun, was largely covered by waters from Earth's primordial ocean. Heat from the Sun diffused through the flux tube's atmosphere, warming these waters and lifting water vapor into the local atmosphere. This vapor condensed as mist on the continents' surfaces, most of which lay at the cooler end of the planet, away from the Sun. Vegetation was abundant, and the surface gravity was 0.66 G, allowing for a taller stature in both flora and fauna. Life was good.

I begin my review of Velikovsky's comments on the period just before the end of the Interim Earth with a single quote:

> *Hebrew mythology assigns to the period preceding Adam's expulsion different geophysical and biological conditions. The sun shone permanently on the Earth, and the Garden of Eden, placed in the East, was, it must be conceived, under perpetual rays of the Dawn. The earth was not watered by rain, but mist ascending from the ground condensed as dew upon the leaves. "The plants looked only to the earth for nourishment." Man was of exceedingly great stature: "The dimensions of man's body were gigantic." His appearance was unlike that of later men: "His body was overlaid with a horny skin." (The Pre-Adamite Age)*

For comparison, here are two more quotes relating to the period immediately after the end of the Interim Earth, describing the initial darkness and then the reemergence of light in a much different form:

In another legend, it is told that the celestial light shone a little in the darkness. And then "the celestial light ceased, to the consternation of Adam." The illumination of the first period never returned. The sky that man was used to seeing never appeared before him again: "The firmament is not the same as the heavens of the first day." The "day" of Genesis, as I have already noted, is said to be equal to a thousand years. (The Pre-Adamite Age)

It was after the fall of man, according to Hebrew tradition, that the sun set for the first time: "The first time Adam witnessed the sinking of the sun, he was seized with anxious fears. All the night he spent in tears. When day began to dawn, he understood that what he had deplored was but the course of nature." It was also then that the seasons began. This is told in the following story: "Adam noticed that the days were growing shorter and feared lest the world be darkened . . . but after the winter solstice he saw that the days grew longer again." (ibid)

From these quotes, I conclude that Adam was alive before and after the end of the Interim Earth, and that he was among the first to witness the Sun's setting and the winter solstice (see above quotes).

Here are four quotes relating to the period after the end of the Interim Earth, including references to the large size of the humans who survived the catastrophe, only to succumb to the 52% increase in gravity that followed. The first line of the first quote below confirms the change in Earth's rotational axis tilt after the end of the Interim Earth, as the resulting day/night cycle and daily temperature changes are necessary preconditions for modern weather patterns, including rainstorms:

The earth also underwent changes: "Independent before, she was hereafter to wait to be watered by the rain from above." The variety of species diminished. Man, according to Hebrew legends, decreased in size; there was a "vast difference between his later and his former state—between his supernatural size then, and his shrunken size now." He also lost his horny skin. The whole of nature altered its ways. (The Pre-Adamite Age)

The traditions of peoples all over the world are quite unanimous in asserting that at an earlier time a race of giants lived on the earth, that most of the race were destroyed in great catastrophes; that they were of cruel nature and were furiously fighting among themselves; that the last of them were exterminated when after a cataclysm a migration of peoples brought the forebears of the peoples of today to their new homelands. (Giants)

Also F. L. Gomara in his Conquista de Mexico, in the chapter about "cinco soles que son edades," wrote: The second sun perished when the sky fell upon the earth; the collapse killed all the people and every living thing; and they say that giants lived in those days, and that to them belong the bones that our Spaniards have found while digging mines and tombs. From their measure and proportion, it seems that those men were twenty hands tall—a very great stature, but quite certain. (ibid)

At the time when the Israelites approached the fields of Bashan in the Transjordan, "only Og king of Bashan" remained of the remnant of the giants (Joshua 13:12 and Deut. 3:11). The other individuals of monstrous size had been annihilated in the meantime. "Behold, his bedstead was a bedstead of iron; is it not in Rabbath of the children of Ammon? nine cubits is the length thereof, and four cubits the breadth of it, after the cubit of a man." The text implies that at the time the book of Deuteronomy was written, the bedstead of Og was still in existence and was a wonder for onlookers. (ibid)

Note the term "hand" as a measurement. Today, this term is used for measuring a horse's height and is equivalent to four inches (10 cm), implying that these Spanish giants were 20 x 4" = 80 inches (about 6 feet 7 inches, or 2 meters) tall, which is not particularly tall by today's standards.

Also, note the term "cubit," an ancient length measurement based on the length of the forearm from the elbow to the fingertips (typically about 18 inches, or 46 cm). Og's bedstead would be 9 x 18" = 162 inches (about 13 feet 6 inches, or 4.1 meters) long and 4 x 18" = 72 inches (about 6 feet, or 1.8 meters) wide. By comparison, a king-size bed today is 80 inches long by 76 inches wide (about 6 feet 6 inches long by about 6 feet

Early Stories and the Ages of Earth

3 inches wide, or 2 meters by 1.9 meters). This suggests that Og could have been twice as tall as a modern human.

Before I begin our discussion of Earth's Ages, I must note that the subject matter is fraught with ambiguity due to the difficulty in assigning specific catastrophes to ages across various cultures. Here's a quote from Velikovsky summarizing the problem:

> *The difference in the magnitude of the catastrophes caused also some nations of antiquity to count six, seven (as most nations), or eight, or nine, or even ten ages; one and the same people, like the Mayas, had traditions of five and seven ages in diverse books of theirs. (Sabbath)*

In my continued description, I'll assume a numbering system for Earth's Ages that considers the time before the end of the Interim Earth as Age Zero since Earth's environment remained mostly unchanged during the human race's evolution. The prominent planet of a period will define that age until the next catastrophe occurs, often due to another planet's actions, which names the following age. Every culture on Earth experienced these changes, and each was forced to reset or recalibrate their particular celestial time-keeping baseline with each age change.

Below is a list of the seven Earth Ages I will be reviewing, showing the chapter number (#), the age name (and number), the approximate time of its beginning years ago (and years from the end of Age Zero), and its duration:

#22: The Age of Uranus (The First Age), about 34,000 years ago
- *(starting at the end of Age Zero and lasting 14,000 years)*

#23: The Age of the Moon (The Second Age), about 20,000 years ago
- *(starting 14,000 years after Age Zero and lasting 8,500 years)*

#24: The Deluge & The Age of Saturn (The Third Age), about 11,500 years ago
- *(starting 22,500 years after Age Zero and lasting 1,500 years)*

#25: The Age of Mercury (The Fourth Age), about 10,000 years ago
- *(starting 24,000 years after Age Zero and lasting 5,800 years)*

#26: The Age of Jupiter (The Fifth Age), about 4,200 years ago
- *(starting 29,800 years after Age Zero and lasting 700 years)*

#27: Exodus & The Age of Venus (The Sixth Age), about 3,500 years ago
- *(starting 30,500 years after Age Zero and lasting 800 years)*

#28 The Age of Mars (The Seventh Age), about 2,700 years ago
- *(starting 31,300 years after Age Zero and lasting 60 years)*

Here are five of Velikovsky's quotes on the subject of Earth Ages:

The Earth underwent re-shaping: six consecutive remoldings. Heaven and Earth were changed in every catastrophe. Six times the Earth was rebuilt—without entire extirpation of life on it, but with major catastrophes. Six ages have passed into the great beyond; this is the seventh creation, the time in which we live. (The Hebrew Cosmogony)

…the ancients also maintained that the successive ages were initiated by planets: Moon, Saturn, Mercury, Jupiter, Venus, Mars. Therefore the sun-ages could also have been called planet ages. (Planet Ages)

Hesiod ascribed the Golden Age to the time when the planet Saturn was ruling, and the Silver and Iron Ages to the time of the planet Jupiter. (ibid)

Vergil, who says that "before Jove's day [i.e., in the Golden Age when Saturn reigned] no tillers subdued the land—even to mark the field or divide it with bounds was unlawful." (ibid)

The idea that the Earth was under the sway of different planets at different ages is also the teaching of the Pythagoreans, the Magi, Gnostic sects, and other secret societies. (ibid)

It is important to recognize that the planets, for which the ages were named, were thought of as gods due to their dramatic impact on Earth and human life. Here are two quotes regarding the deification of the planets:

The constellations of the sky took only a minor and incidental part in the mythology of the ancient peoples. The planets were the major gods, and they ruled the universe. (Deification of the Planets)

…the same process of identification of major gods with the planets can be found in the religions of peoples in all parts of the world. The planets were not affiliated with the gods, or symbols of the gods—they were the gods. (ibid)

Surprised? The planets were not entities associated with gods…they *were* the gods.

Five Friends

Chapter 22

The Age of Uranus (The First Age)

What prompted 500 generations of planet gods storytelling?

Thirty-four thousand years ago, the event that ended the Interim Earth (the destruction of the flux tube) initiated a chaotic period for the solar system. Four new gas giant planets—Saturn, Jupiter, Uranus, and Neptune (the products of Proto-Saturn's breakup during the event)—and two rocky planets, Earth and Mars, which survived the event but were significantly impacted, orbited the Sun. Major debris from the breakup of Proto-Venus shortly after the event, including Earth's Moon, Mercury, Venus, and Phaeton, were still in various orbits around the gas giants. Though perhaps visible to humans on Earth, these bodies had not yet assumed a prominent role in the Earth Ages. As this narrative unfolds, I will describe how three of these planet-sized objects—Earth's Moon, Mercury, and Venus—caused Earth-shaking catastrophes, earning their places in the list of Earth Ages. The fourth, Phaeton, would play a prominent role in the Age of Saturn before being swallowed by the Sun.

For 14,000 years, Saturn, Jupiter, Uranus, and Neptune moved erratically, with their orbits crossing and recrossing each other as they constantly changed position and relative prominence in Earth's sky. Once the dust and clouds had cleared from Earth's final shape shift, the surviving inhabitants watched these giants seemingly battle for dominance. Neptune had already retreated to a distance that removed it from contention, leaving Uranus, Saturn, and Jupiter as objects of worship for Earth's people. For nearly 500 generations, stories of the planet gods were told and retold, with no clear victor in the Ages-of-the-Earth saga.

Eventually, however, Uranus took action that made it stand out from the other, larger planets, Saturn and Jupiter. The details of Uranus's distinguishing actions will be discussed in the next chapter, Chapter 23. Here are two quotes about Uranus:

"The seven planets of the ancients comprised the Sun, the Moon, Mercury, Venus, Mars, Jupiter, and Saturn. However, the ancients' religions and mythology speak for their knowledge of Uranus; the dynasty of gods had Uranus followed by Saturn, and the latter by Jupiter. In the clear skies of Babylonia, the planet Uranus could have been observed by the unaided eye; but since it was known as a deposed deity, it would seem that at some later time, the planet lost much of its brightness." *(Uranus)*

"It is quite possible that the planet Uranus is the very planet known by this name to the ancients. The Age of Uranus preceded the Age of Saturn; it came to an end with the 'removal' of Uranus by Saturn. Saturn is said to have emasculated his father Uranus." *(ibid)*

Though Uranus is not often included in other lists of Earth Ages or in many of Earth's cultures, it earned its place as the god of the first Earth Age in my numbering system by initiating the sequence that led to the second Earth Age, a time before the Moon appeared in Earth's sky.

Chapter 23

The Age of the Moon
(The Second Age)

The period when the Earth was moonless is probably the most remote recollection of mankind.

About 20,000 years ago, approximately 14,000 years after the end of the Interim Earth, the elliptical orbits of Uranus and Saturn aligned so closely that they approached each other on one of their passes around the Sun. Since the mass of Saturn is much larger than that of Uranus, this interaction forced Uranus into a larger orbit, sending it much farther from the Sun and reducing its apparent size to observers on Earth. In the process, the largest moon-sized object captured by Uranus from the debris of Proto-Venus's breakup was stripped away and sent into an independent orbit around the Sun. This moon-sized object then started on a near-collision course with Earth, still retaining most of the electrical charge it held when it had been part of Proto-Venus.

As this errant moon approached Earth's plasmasphere with its dramatically different voltage, the oppositely charged electrical fields of both bodies interacted, causing violent bolts of electricity to jump between them. These electrical discharges and other energy-dissipating forces allowed Earth to capture this object as its own permanent Moon. After many orbits, Earth's new Moon settled into a near-circular orbit much closer to Earth than its present distance. The gravitational and electrical interaction during the approach and capture caused immense ocean tidal waves, raised mountains to new heights, and fractured the Earth's surface, allowing lava to spill forth.

Humans feared and revered this new god in their heavens, holding it in higher esteem than the Sun. It had become their sky's largest and brightest object—it took its obvious place as their supreme deity. Here are two quotes supporting our discussion of mankind's worship of the Moon as a god:

"The sun was of smaller importance than the moon in the eyes of the Babylonian astrologers." (The Worship of the Moon)

The Assyrians and the Chaldeans referred to the time of the Moon god as the oldest period in the memory of the people. Before other planetary gods came to dominate the world ages, the Moon was the supreme deity. (ibid.)

Below are five quotes that serve to authenticate that there was a time period within the memory of man when there was no moon overhead:

Democritus and Anaxagoras (Greek philosophers from around 400-500 years BCE) taught that there was a time when the Earth was without the Moon. Aristotle wrote that Arcadia in Greece, before being inhabited by the Hellenes, had a population of Pelasgians, and that these aborigines occupied the land before there was a moon in the sky above the Earth; for this reason, they were called Proselenes ("selene" is Greek for *moon*).

Apollonius of Rhodes mentioned the time "when not all the orbs were yet in the heavens, before the Danai and Deukalion races came into existence, and only the Arcadians lived, of whom it is said that they dwelt on mountains and fed on acorns, before there was a moon." (ibid.)

The memory of a world without our Moon lives in oral tradition among the Indians. The Indians of the Bogota highlands in the eastern Cordilleras of Colombia relate some of their tribal reminiscences to the time before there was a moon. "In the earliest times, when the moon was not yet in the heavens," say the tribesmen of Chibchas. (ibid.)

The Age of the Moon (The Second Age)

...mankind on both sides of the Atlantic preserved the memory of a time when the Earth was without the moon. (ibid.)

We have seen that the traditions of diverse peoples offer corroborative testimony to the effect that in a very early age, but still in the memory of mankind, no moon accompanied the Earth. (ibid.)

My contention that the Moon (when it was first captured) was orbiting closer to the Earth than it is today is supported by these two quotes:

"Many traditions persist that at some time in the past the Moon was much brighter than it is now and larger in appearance than the Sun." (A Brighter Moon)

In fact, the Babylonian astronomers computed the visible diameter of the Sun as only two-thirds of the visible diameter of the Moon, which makes a relation of four to nine for the illuminating surfaces. This measure surprised modern scholars, who are aware of the exactness of the measurements made by the Babylonian astronomers and who reason that during eclipses, one can easily observe the approximate equality of the visible disks. (ibid.)

The three quotes below discuss the time after the Age of the Moon had passed, as the Moon was shifted in its orbit but never removed from Earth by the influence of other planets as they began their respective ages:

Since the time the Moon began to accompany the Earth, it underwent the influence of contacts with comets and planets that passed near the Earth in subsequent ages. The mass of the Moon, being less than that of the Earth, must have suffered greater disturbances in cosmic contacts. During these contacts, the Moon was not carried away; this is due to the fact that no body more powerful than Earth came close enough to remove the Moon permanently. However, in these contacts, the Moon was repeatedly shifted from one orbit to another. (The Earth Without the Moon)

Whatever atmosphere it may have had…was pulled away by Earth, by other contacting bodies, or dissipated in some other way. (ibid.)

The variations in the position of the Moon can be read in the variations in the length of the month. The length of the month repeatedly changed in subsequent catastrophic events—and for this, there exists a large amount of supporting evidence. In these later occurrences, the Moon played a passive role, and Zeus in the Iliad advised it (Aphrodite) to stay out of the battle in which Athene and Ares (Venus and Mars) were the main contestants. (ibid.)

Chapter 24

The Deluge and
The Age of Saturn (The Third Age)

When you hike down the bluff to Lake Michigan, you take a trip back in time. At first, the beach looks as if blanketed by multi-hued stones and rocks—many of these are actually fossils. Fossils are nature's way of revealing evidence of prehistoric organisms, and the evidence in this region comes from a time long before the Great Lakes formed—before many significant events impacted Earth, including The Flood.

About 11,500 years ago, some 8,500 years after Earth acquired its Moon, and approximately 22,500 years after the end of the Interim Earth, the final shock wave from the stellar fissioning of Geminga swept across the still chaotic solar system. The impact disturbed the orbits of Saturn and Jupiter enough that they passed in close proximity, allowing Jupiter's massive gravity to pull a portion of Saturn's atmosphere into space, creating a nova-like explosion. The electrically charged ejecta contained clouds of gases (originally part of the flux tube's atmosphere) and spores from Proto-Venus's Earth-like flora, decimated at the end of the Interim Earth.

Some debris from this event included hypervelocity iron particles that impacted broad areas of Earth's northern hemisphere. These are the iron particles that the authors of *Cycle of Cosmic Catastrophes* found in abundance, mixed in the Clovis Layer at numerous sites in North America, as I described in Chapter 20.

The explosion also pushed two planet-sized satellites from their orbits around Saturn. The one that would become the planet Mercury was ejected into an elliptical orbit with a perihelion close to the Sun. The one the Sumerians called *Phaeton* (or *Marduk*) was freed from the vicinity of Saturn and Jupiter and sent on a course headed directly toward Earth. Both objects streamed tails of ejecta behind them, giving them a comet-like appearance. As Phaeton sped earthward, it was observed passing close enough to Mars for at least one electrical discharge between them to occur. Several icy remnants of Proto-Venus' ocean accompanied Phaeton on its journey.

For seven days, the light from this explosion illuminated Earth's entire sky, both day and night. On the eighth day, the sky darkened when part of Saturn's ejected atmosphere contacted Earth, bringing clouds, lightning, and a hot, torrential rain that obscured further view of Phaeton's approach. Some rocky and icy debris from Saturn's atmosphere, orbiting Phaeton on its approach to Earth's North Pole, were the first objects to strike. The lightweight icy debris entered Earth's atmosphere as meteoric objects, smashing into the surface and sending fractured shards in a secondary impact pattern over a wide area. Among the artifacts of the primary impacts seen today are Hudson's Bay and Lake Michigan, and secondary impacts include the Carolina Bays.

As Phaeton approached Earth from the north, the gravitational and static electric forces between them pulled the oceans toward the North Pole, piling seawater high enough to cover mountains. These forces distorted Earth's crust, so that the area directly facing Phaeton bulged upward, pulling Earth into a slightly ellipsoidal shape. As Phaeton drew closer, the bulge increased, and when Phaeton passed very close overhead, the bulging portion of the crust was dragged along. This caused Earth's crust to shift over its massive magma core like a skin sliding over a sphere. The nearly spherical shape of Earth facilitated this crustal movement. Friction between the crust and core heated magma to extreme temperatures, causing it to spew through newly formed cracks in the crust.

The Deluge and The Age of Saturn (The Third Age)

At this same moment, an electrical discharge between Earth and Phaeton relieved the electrostatic tension, allowing the waters to cascade dramatically across the land, carrying boulders, trees, and animals, piling debris and cramming remains of creatures into fissures and caves on high mountains.

When Phaeton pulled away, the crust stopped sliding over the molten core, coming to rest with land that had been nearly a thousand miles farther south now near the North Pole. Earth's spin axis had not tilted with respect to the plane of the ecliptic, but every location on Earth had new longitude and latitude coordinates. Many areas in temperate climates moved closer to the poles, shifting into polar climates. Many animals killed by the floodwaters and debris waves left their bodies flash-frozen in mud as temperatures plunged. Countless fossils have been found.

Earth's surface was left fractured with long rifts; mountain ranges were lifted to new heights; level land masses tilted, causing rivers to run in reverse and lakes to drain; volcanoes erupted; lava flowed onto Earth's surface; and massive extinctions occurred.

For years after the deluge, Earth's atmosphere was choked with smoke and ash. The Sun was totally obscured, and the entire world was cloaked in darkness. When the sky finally cleared and survivors looked to the heavens, they found nothing as it had been before. Saturn, last seen colliding with Jupiter before the deluge, was no longer in its place in the heavens, while Jupiter appeared much more prominent. When Saturn was eventually detected, it appeared much smaller, moving more slowly and surrounded by rings. The Moon was in a larger orbit around Earth, appearing smaller, with a changed orbital period. The Sun, stars, Moon, and planets traveled in different paths across the sky, requiring new calendars and sky charts. A new Age had dawned.

Survivors found that land masses had shifted. Continents had changed shape and position. Areas once dry land were now submerged, and previously submerged areas had been lifted. The ocean water was very salty. Many landmasses were covered with shallow, toxic algae pools that covered the bodies of decimated animals.

Golden Age of Saturn

The catastrophe Saturn visited upon Earth was the most damaging since the end of the Interim Earth. This was the biblical Noah's Flood, the Deluge, and would prove to be the worst of the remaining Age-defining catastrophes, ushering in the Age of Saturn. This age is not to be confused with the Golden Age of Saturn, which refers to the last 10,000 years of the Interim Earth, the time known as the biblical Garden of Eden. I am including here Velikovsky's thoughts on this complex subject.

Below are two quotes from passages written by the ancients that describe the Golden Age of Saturn:

Hesiod tells of … "A golden race of mortal men who lived in the time of Kronos when he was reining in heaven. And they lived like gods without sorrow of heart, remote and free from toil: miserable age rested not on them... The fruitful earth unforced bore them fruit abundantly and without stint. They dwelt in ease and peace upon their lands with many good things.". (Saturn's Golden Age)

Similarly, Ovid writes in the sixth book of his *Metamorphoses*: "In the beginning was the Golden Age, when men of their own accord, without threat of punishment, without laws, maintained good faith and did what was right... The earth itself, without compulsion, untouched by the hoe, unfurrowed by any share, produced all things spontaneously... It was a season of everlasting spring.". (ibid.)

Compare the last line of the second quote above ("It was a season of everlasting spring") to the line from the first quote in Chapter 21 that reads: "The sun shone permanently on the Earth, and the Garden of Eden, placed in the East, was, it must be conceived, under perpetual rays of the Dawn." Notice the similarity. The only Earth orientation that would provide an "everlasting spring" after the end of the Interim Earth would be if Earth's axis were perpendicular to the plane of the ecliptic.

The Deluge and The Age of Saturn (The Third Age)

But from two lines in the third quote, it can be seen that this was not the case. From the line "… that the sun set for the first time," it is clear that this quote describes a time immediately after the end of the Interim Earth. The line "It was also then that the seasons began" implies the axis was close to its present tilt of 67 degrees from the plane of the ecliptic, allowing seasons to exist. The conclusion can only be that the Golden Age of Saturn occurred before the end of the Interim Earth, as that is the only time Earth saw a perpetual spring.

In the three quotes below, Velikovsky ponders the discrepancy between the term the Golden Age of Saturn (Kronos) and the naming of the period after the deluge simply the "Age of Saturn."

"The dominance of Saturn at some remote period in the history of the life of the peoples on Earth was of such pronounced and all-pervading character that the question arises whether the adventures of the planet going through many exploits could by itself be the full cause of the worship of the planet and the naming of the Golden Age 'the Age of Kronos' (Saturn). Saturn exploded and caused the Earth to go through the greatest of its historical catastrophes, and this was completely sufficient to make of Saturn the supreme deity; but it appears that the Age of Saturn is a name for the epoch before the Deluge; after the Deluge Saturn, dismembered, almost ceased to exist as a planetary body and when at length it was reconstituted it was fettered by rings, and was far from being the dominant celestial body that would behoove it as the supreme deity of the epoch. The 'Age of Kronos' is so glorious an age that it is hardly thinkable to connect it with the period after the Deluge. The wailing for Adonis, Tammuz of the Babylonians, or Osiris of the Egyptians, deplored the end of its dominance, not the beginning of it.". (Saturn's Golden Age)

"Then why was Saturn the supreme deity by whose name the great and glorious age before the Deluge was named? Because it removed Uranus from its role of chief deity, and to the onlookers on Earth, emasculated him? (ibid.)

"The Golden Age of Saturn or Kronos came to its end when the supreme god of that period, the planet Saturn, was broken up. The Age of Kronos was not the earliest age of which man retained some, however dim, memories—but farther into the past the dimness amounts almost to darkness."

It is in the line "The Golden Age of Saturn or Kronos came to its end when the supreme god of that period, the planet Saturn, was broken up" that I find the clincher for placing the Golden Age of Saturn at the end of the Interim Earth: the breakup of Proto-Saturn as one of the consequences of the loss of the flux tube that ended that era, and the ancients gave both Proto-Saturn and Saturn the same name.

Predictions

Ancient civilizations watched the movements of the planets and from their observations could determine a close encounter between the orbits of Saturn and Jupiter. The quote below describes one of these predictions:

"According to Hebrew legends, the Deluge and its time had already been predicted by Enoch, and even more ancient generations were said to have erected tablets with calendric and astronomical calculations predicting the catastrophe…This might have been the knowledge of months, of years, and of the periods of comets that the remote generations had acquired." (Astronomical Knowledge Before the Deluge)

Worship

The prominence of Proto-Saturn in the final days of the Interim Earth, and the actions of his namesake, the planet Saturn, for tens of thousands of years after that era ended, earned him a place in the minds of the ancients as their supreme god. The two quotes below address the time of Saturn's dominance as the supreme deity and the eventual surrender of that title:

"Saturn, so active in the cosmic changes, was regarded by all mankind as the supreme god.". (The Worship of Saturn)

The Deluge and The Age of Saturn (The Third Age)

"The different names for God in the Bible reflect the process of going through the many ages in which one planet superseded another and was again superseded by the next one in the celestial war. El was the name of Saturn; Adonis of the Syrians, the bewailed deity, was also, like Osiris, the planet Saturn; but in the period of the contest between the two major planets, Jupiter and Saturn, the appellative of the dual gods became Adonai, which means 'my lords'; then, with the victory of Jupiter, it came to be applied to him alone." (ibid.)

Two Stars

The predictions of the ancients that Jupiter and Saturn would interact came true as the two planets moved closer together on their intersecting orbits, allowing the combined effect of their electrical and gravitational fields to dislodge Mercury, Phaeton, and other Proto-Venus debris from their orbits around Saturn. The quote below bears witness to this event:

"The passage in the Talmud that makes the planet Khima responsible for the Deluge means: 'Two stars erupted from the planet Saturn and caused the Deluge.'" (Khima)

Nova

As Saturn and Jupiter approached even closer, their interaction resulted in a nova-like explosion, as described in the two quotes below:

"As told, the picture that emerges from comparative folklore and mythology presents Saturn and Jupiter in vigorous interactions. Suppose that these two bodies approached each other rather closely at one time, causing violent perturbations and huge tidal effects in each other's atmospheres. Their mutual disturbance led to a stellar explosion, or nova. As we have seen, a nova is thought to result from an instability in a star, generated by a sudden influx of matter, usually derived from its companion in a binary system. (Saturn and Jupiter)

"On becoming a nova, it ejected filaments in all directions and the solar system became illuminated as if by a hundred suns. It subsided rather quickly and retreated into far-away regions." (The Worship of Saturn)

Velikovsky suggests a possible date for the nova-like explosion of Saturn as about 10,000 years ago. This is close to the date suggested in *Cataclysm* of 11,500 years ago that I have been using for my thesis. He writes:

"All that we have considered up to now indicates that Saturn once exploded in a nova-like burst of light. The date of this event I would be hard-put to specify, even approximately, but possibly it took place about ten thousand years ago. The solar system and reaches beyond it were illuminated by the exploded star, and in a matter of a week the Earth was enveloped in waters of Saturnian origin. (Star of the Sun)

Seven Days of Lights

The nova-like explosion of Saturn sent a flood of light through the solar system that was brighter than the Sun and lasted for seven days, as described in the following four quotes:

"The light of the seven days served the population of the world as a warning of some extraordinary events." (The Light of the Seven Days)

"…the blinding light preceding the Deluge by seven days is an interesting and important detail. The world was in a dazzling light, sevenfold stronger than the light of the sun; the light was so strong and so brilliant day and night alike, that the sun was entirely overpowered by it; and in the days of Isaiah, thousands of years later, the memory of the light of the seven days was vivid in tradition, so that the prophet could refer to it in desiring to describe the solar light of the messianic age." (ibid.)

The Deluge and The Age of Saturn (The Third Age)

"The Deluge and the seven days of brilliant light immediately preceding it were a universal experience, and they left indelible memories. Many of the religious rites and observances of all creeds go back to these events of the past in which the celestial gods Saturn and Jupiter were the main participants. Among the most ancient of all such observances were festivals of light of seven days' duration, held in honor of Saturn. The 'seven days of light' just before the Deluge overwhelmed the Earth are recreated in these feasts." (Festivals of Light)

"During the 'seven days' when the world was flooded by sheets of light, and terrifying signs and commotion filled the heavens, 'the Holy One . . . reversed the order of nature, the sun rising in the west and setting in the east.'" (The Deluge in Rabbinical Sources)

A quote above describes the Sun as "rising in the west and setting in the east." To explain this seemingly impossible observation, I have developed the following simple rationale: Suppose that the Rabbinical sources who are quoted here were observing the nighttime sky around three o'clock in the morning at a time when the Sun, Earth, and Saturn were in alignment. Then, after watching Saturn set in the west, suppose they turned their attention to the east in anticipation of the coming sunrise. If the first flare-up of Saturn's nova-like explosion occurred just after Saturn dropped below the western horizon, its intense light would have flooded the western sky, causing all who were there assembled to turn back to the west in wonder at what appeared to be the early light of dawn occurring in the western sky instead of in the east.

By the time the Sun actually arose in the east, its light could not be distinguished from the nova light, as explained above, "the light was so strong and so brilliant day and night alike, that the sun was entirely overpowered by it." Then, seven days later, after that day's actual sunset had gone unnoticed, if the nova flare died out just before Saturn rose, it would have left its last light in the eastern sky, giving the impression that the Sun was setting in the east.

The Flood

The following eleven quotes deal with several aspects of the actual flood, including when it started, its universality, where the water came from, its depth, darkness, biblical references, the temperature and salinity of the water, and changes in the Earth's atmosphere:

"The tradition of a universal deluge is told by all ancient civilizations, and also by races that never reached the ability to express themselves in the written symbols of a language. It is found all over the world, on all continents, on the islands of the Pacific and Atlantic, everywhere. Usually, it is explained as a local experience carried from race to race by word of mouth. The work of collating such material has repeatedly been done, and it would only fatigue the reader were I to repeat these stories as told in all parts of the world, even in places never visited by missionaries." (Deluge)

"…but all agree that the earth was covered to the mountaintops by the water of the deluge coming from above, and that only a few human beings escaped death in the flood. The stories are often accompanied by details about a simultaneous cleavage of the earth." (ibid.)

"But the climatologists make it very clear that even should the entire water content of the atmosphere pour down as rain, the resulting flood could not have covered even the lowland slopes, far less the peaks of the mountains, as all accounts insist that this deluge did." (ibid.)

"Following the 'seven days' when the world appeared to be ablaze in 'the radiance of a thousand suns' the Deluge started." (Arrival of the Waters)

"It has a Babylonian counterpart in the title of Tammuz as *bel girsu*: 'lord of the flood.' The nova blazed terrifically, but soon the light became diffused, the shadows grew ever dimmer, the world

The Deluge and The Age of Saturn (The Third Age)

that was all splendor and light turned gloomier and gloomier; the outpouring waters grew ever thicker; the clouds of dust darkened ever more the sky, and finally the drama of what was taking place on earth went on in darkness." (ibid.)

"The Flood was caused by waters pouring from above, but also by waters drawn up from the ground. 'All the fountains of the great deep were broken up, and all the windows of heaven were opened.' The waters that came from the sky were heated. Many passages in the rabbinical literature refer to the heated water." (The Deluge in Rabbinical Sources)

"There must have been many Noahs, and the Midrashim also say so—but probably none of them escaped with his boat the outrages of nature. Possibly, in some caves high in the mountains, in far-separated regions of the earth, human beings survived the Deluge; but hardly any vessel or ark." (ibid.)

"The conflict between the larger planets resulted in long-stretched filaments ejected by a disturbed Saturn to cross the Earth's orbit. The hydrogen of the planet combined with the oxygen of the terrestrial atmosphere in electrical discharges and turned into water…There are definite indications of a drastic drop in the atmospheric oxygen at the time of the Deluge—for instance, the survivors of the catastrophe are said in many sources to have been unable to light fires…The consumption of the oxygen in the air by its conversion into water could not fail to have a marked effect upon all that breathes. The animal life that survived needed to accommodate itself to the changed conditions…According to rabbinical sources, before the Deluge man was vegetarian; but the post-diluvian population did not continue the vegetarian habits of the 'sinful' population of the earth. The Talmud and the Midrashim narrate that after the Deluge a carnivorous instinct was awakened in animal and man, and everyone had the impulse to bite." (Hydrogen and Oxygen)

"In the year of the creation of the world 1656, after Noah had attained the age of 600 years, three days before the death of Methuselah, a comet appeared in the constellation Pisces, was seen by the entire world as it traversed the twelve signs of the zodiac in the space of a month; on the sixteenth of April it again disappeared. After this the Deluge immediately followed, in which all creatures which live on earth and creep on the ground were drowned, with the exception of Noah and the rest of the creatures that had gone with him into the ark. About these things is written in Genesis, chapter 7." (Deluge and Comet)

"But when a world cycle perishes by water…there arises a cycle-destroying great cloud of salt water. At first it rains with a very fine rain which gradually increases to great torrents which fill one hundred thousand times ten million worlds, and then the mountain peaks of the earth become flooded with saltish water, and hidden from view. And the water is buoyed up on all sides by the wind, and rises upward from the earth until it engulfs the heavens." (The Origin of the Oceans)

"The knowledge that the water of the oceans came for the most part from Saturn and that the waters were salty was combined by the Greeks into a metaphor which has the sea being the 'tear of Kronos.'" (ibid.)

Saturn's Fate

The following three quotes describe the fate of Saturn after the flood:

"Peoples that remembered early tragedies enacted in the sky by the heavenly bodies asserted that Jupiter drove Saturn away from its place in the sky. Before Jupiter (Zeus) became the chief god, Saturn (Kronos) occupied the celestial throne. In all ancient religions, the dominion passes from Saturn to Jupiter. In Greek mythology, Kronos is presented as the father and Zeus as his son who dethrones him. Kronos devours some of his children.

The Deluge and The Age of Saturn (The Third Age)

After this act Zeus overpowers his father, puts him in chains, and drives him from his royal station in the sky. In Egyptian folklore or religion the participants of the drama are said to be Osiris-Saturn, brother and husband of Isis-Jupiter." (The Worship of Saturn)

"For a time Saturn disappeared, driven away by Jupiter, and when it reappeared it was no longer the same planet: it moved very slowly. The disappearance of the planet Saturn in the 'nether world' became the theme of many religious observances, comprising liturgies, mystery plays, lamentations, and fasts. When Osiris was seen again in the sky, though greatly diminished, the people were frenzied by the return of Osiris from death; nevertheless, he became king of the netherworld." (ibid.)

"In the Egyptian way of seeing the celestial drama, Isis (Jupiter), the spouse of Osiris (Saturn) wrapped him in swathings. Osiris was known as 'the swathed'—the way the dead came to be dressed for their journey to the world of the dead, over which Osiris reigns." (ibid.)

Changes to the Earth

Many changes to flora, fauna, and the land were wrought by the Saturn/Phaeton encounter:

"Like the former catastrophe of the fall of man, this catastrophe of the Deluge, according to the Hebrew cosmogony, changed the nature of herb, animal, and man. The prosperity of the time before the great flood was gone, never to return; the world lay in ruins. The earth was changed; even the sky was not the same." (The Deluge in Rabbinical Sources)

"The areas which are now the shores of the Mediterranean were the shores of an open ocean—or so one may conclude from the following statement: 'Before the birth of Noah, the sea was in the habit of transgressing its bounds twice daily, morning and evening. Afterwards, it kept within its confines.'" (ibid.)

Seeds

As the Earth recovered from the Deluge, the flora grew abundantly with new and strange species appearing for the first time. The following writings explore this phenomenon and describe how an exploding planet (think Proto-Venus) may have been involved:

"Saturn was called 'the god of seeds' or 'of sowing,' also 'the lord of the fieldfruits.'" (Saturn the God of Seeds)

"A Deluge destroying much faunal life must have caused dissemination of plants: in many places, new forms of vegetation must have sprouted from the rich soil fertilized by lava and mud; seeds were carried from all parts of the globe and in many instances, because of the change in climate, they were able to grow in new surroundings. The axis of the earth was displaced, the orbit changed, the speed of rotation altered, the conditions of irrigation became different, the composition of the atmosphere was not the same—entirely new conditions of growth prevailed." (ibid.)

"Ovid thus describes the exuberant growth of vegetation following the Flood: 'After the old moisture remaining from the Flood had grown warm from the rays of the sun, the slime of the wet marshes swelled with heat, and the fertile seeds of life, nourished in that life-giving soil, as in a mother's womb, grew, and in time took on some special form.' 'When, therefore, the earth, covered with mud from the recent Flood, became heated up by the hot and genial rays of the sun, she brought forth innumerable forms of life, in part of ancient shapes, and in part creatures new and strange.'" (ibid)

"The innumerable new forms of life in the animal and plant kingdoms following the Deluge could have been solely a result of multiple mutations…Although this seems a sufficient explanation of why and how Saturn came to be credited with the work of dissemination and mutation, the mention of another possibility should not be omitted." (ibid.)

"If it is true that the Earth passed through the gases exploded from Saturn, it should not be entirely excluded that germs were carried together with meteorites and gases and thus reached the Earth." (ibid.)

"The scholarly world in recent years has occupied itself with the idea that microorganisms—living cells or spores—can reach the Earth from interstellar spaces, carried along by the pressure of light rays. The explosion of a planet is a more likely method of carrying seeds and spores through interplanetary spaces…The new forms of life could be the result of mutations, a subject I have discussed in Earth in Upheaval. But the possibility that seeds were carried away from an exploding planet cannot be dismissed either." (ibid.)

Rings

Saturn is known for its rings. The following quotes discuss the appearance of Saturn swathed in rings after the Deluge and confirm that the ancients of many cultures could see them with the unaided eye due to Saturn's close proximity to Earth at the time:

"In the Egyptian legend, Isis (Jupiter) swathes Osiris (Saturn). The Egyptian appellative for Osiris was 'the swathed.'" (The Rings of Saturn)

"The rings of Saturn were known also to the aborigines of America before Columbus discovered the land; this means also before the telescope was invented at the beginning of the seventeenth century. An ancient engraved wooden panel from Mexico shows the family of the planets: one of them is Saturn, easily recognizable by its rings." (ibid.)

"…the ancient Maoris of New Zealand knew about her rings—for there is evidence that they did have a Saturnian ring legend long before the days of Galileo." (ibid.)

"…the knowledge of the ancients about the rings of Saturn could have been acquired because of better visibility: in other words, at some time in the past Saturn and Earth appear to have been closer to one another." (ibid.)

For the next 5,000 years, the Earth was at peace as it recovered and blossomed after the Deluge. Though the Age was named for the now-deposed Saturn, and with Jupiter dominating the nighttime sky, a tiny spot of light orbiting the Sun was about to raise havoc for the Earth.

Chapter 25

The Age of Mercury (The Fourth Age)

About 10,000 years ago, the descendants of the giants who survived the Deluge built the city of Babylon on the plains of Mesopotamia. The surrounding land was fertile from the soil deposited by the great flood, allowing the city to prosper and develop into a cultural center. Wise men among the inhabitants constantly watched the skies to detect any warnings of a possible repeat of the flood that had occurred 1,500 years earlier and was still vivid in their societal memory.

They carefully observed a wandering star that their ancestors had identified as having been torn loose from its orbit around Saturn during its encounter with Jupiter just before the Deluge. The Babylonians were a strong and industrious people, so when the wise men warned them that this rogue star was on a path likely to encounter Earth, they decided to build a tall structure to act as a refuge should another flood befall them.

They built a tower of fired brick and stone in the middle of the plain that was a wonder to all who saw it. As predicted, the rogue star, which they called Mercury, approached Earth. It came close enough for the magnetospheres of both planets to make contact, allowing violent electrical discharges to occur between the two bodies. One strong bolt hit the top of the tower as though it were a lightning rod, destroying the tower and subjecting the people of Babylon to strong electrical currents that caused momentary amnesia and temporary confusion of speech. The atmospheric disturbances caused violent hurricane winds to blow, destroying other structures, but did not bring the flood they had feared. After the encounter, Mercury changed its path and entered an orbit around the Sun closer than Earth's. The survivors of this latest catastrophe could see Mercury thereafter as a morning or evening star.

The eleven quotes below describe the advent of the Age of Mercury and the consequences of its encounter with Earth, including comments on the physical destruction, the mental effects on the inhabitants, the timing of the event (which agrees with the timetable described in this book), the spread of the story, the combatants (giants vs. Mercury), and the worship of this new god:

> "It appears that after the Flood the plain of Mesopotamia became one of the few cultural centers of the world. Another flood would have caused the utter destruction of the human race, and this was feared because the memory of the Flood a few centuries earlier was very vivid. Observations of the movements of the heavenly bodies may have provided a warning of a new catastrophe and large structures were built for refuge. But when the event came, the structures were overwhelmed and destroyed by hurricanes and powerful electrical discharges." (The Confusion of Languages)

> "The Babylonian account, as transmitted by Abydenus, tells that once men 'built a high tower where now is Babylon, and when it was already close to heaven, the gods sent winds and ruined the entire scheme... and men, having till then been all of the same speech, received [now] from the gods many languages.'" (ibid.)

> "Other accounts give the impression that a strong electrical discharge—possibly from an overcharged ionosphere—found a contact body in the high structure. According to a tradition known to the twelfth-century traveler Benjamin of Tudela, 'fire from heaven fell in the midst of the tower and broke it asunder.'" (ibid.)

> "The Tower of Babel story was found in the most remote parts of the world prior to the arrival of missionaries in those places, thus before the Biblical account became known to the aborigines." (ibid.)

The Age of Mercury (The Fourth Age)

"In the rabbinical concept of the seven earths, molded one out of another in successive catastrophes, the generation which built the Tower of Babel inhabited the fourth earth; but it goes on to the fifth earth where the men become oblivious of their origin and home: those who built the Tower of Babel are told to forget their language. This generation is called 'the people who lost their memory.' The earth which they inhabited was 'the fifth earth, that of oblivion (Neshiah).'" (ibid.)

"The characteristic of this catastrophe was its influence upon the mental, or mnemonic, capacity of the peoples. The description of it, as told by many tribes and peoples, if it contains authentic features, arouses the surmise that the earth underwent an electromagnetic disturbance, and that the human race experienced something that in modern terms seems like a consequence of a deep electrical shock." (ibid.)

"The application of electrical current to the head of a human being often results in a partial loss of memory; also a loss of speech may be induced by the application of electrodes to specific areas of the brain." (ibid.)

"The earliest allusion to these events is in Homer's Odyssey... Homer ascribes the destruction of the giants to Apollo. Pliny ... and Macrobius... identified Apollo with the planet Mercury. Apuleius wrote ... that Mercury and Apollo were alternate names for 'Stilbon,' the planet Mercury. Hesiod described the battle with the giants as an immense catastrophe involving the earth and heaven alike." (ibid.)

"The claim, therefore, is that Mercury has traveled on its present orbit for only some five or six thousand years. This view conflicts with both the nebular and the tidal theories of the origin of the planetary family, and with the assumption that the planets have occupied the same orbits for billions of years." (Mercury)

"'Above all they worship Mercury, and count it no sin to win his favor on certain days by human sacrifices.' Odin was the head of the Nordic pantheon.". (ibid.)

"We worship the gods of our fathers, that is, Jupiter, Saturn, and the rest of those that rule the world, but most of all [we worship] Mercury, whom in our language we call Voden… Of Odin it was said: 'He spoke so well and so smoothly that all who heard him believed all he said was true.'" (ibid.)

For 700 years, Mercury was the supreme deity over the Earth. Life on the planet recovered quickly from the encounter with Mercury, but the peace and tranquility were not destined to last. Jupiter, still the most prominent object in the ancient sky, was about to roil the pot once again.

Chapter 26

The Age of Jupiter (The Fifth Age)

Can you imagine looking out the window of your spacecraft and seeing a cloud of dust, only to realize it was actually gold dust? "Jumping Jupiter!" you'd exclaim, and you'd be right.

About 4,200 years ago, civilization was advancing, and the Bronze Age was in full swing. The First Dynasty of ancient Egypt had been founded. Stonehenge had been built in what is now Wiltshire, England. The populous cities of Sodom, Gomorrah, Admah, Zeboiim, and Zoar had been built on the plain of Jordan, with well-watered land all around. Jupiter, the dominant planet in the sky, was still recovering from its expulsion of Saturn to the outer reaches of the solar system and the deflection of Mercury to an encounter with Earth 5,800 years earlier at the time of the Tower of Babylon.

Venus was being held in an irregular and unstable orbit around Jupiter by electromagnetic and gravitational forces. On one close orbital pass, an electrical discharge between the two changed Venus's trajectory to a path that allowed it to escape from its orbit around Jupiter. The ensuing disturbance of Jupiter's atmosphere resulted in the ejection of a highly charged tendril of primordial debris, left over from the destruction of Proto-Venus, that had been collected by the newly formed planet Jupiter almost 30,000 years earlier. Some of that debris was gold dust.

In Chapter 13, I discussed the effects of the Earth's shape change that followed the catastrophe that ended the Early Earth. In Figure 7, View E,

I showed a fracture in the Earth's crust that parallels the west coast of the African continent. That crack appeared in response to the gradual circumferential stretching that took place during the Earth's ellipsoidal shape change from a length-to-diameter ratio of approximately 1.6 to 1 to approximately 1.4 to 1. With Earth's final shape change at the end of the Interim Earth, that fracture zone was stressed for a second time. During the Deluge, 7,500 years before the time of our present discussion, the entire crust was shifted on the Earth's molten core, moving the plain of Jordan closer to the equator, where it was about to become ground zero for the most powerful thunderbolt strike the world had ever seen.

As the Earth and Jupiter came into orbital alignment in opposition (with both planets aligned on the same side of the Sun), Jupiter's filament of charged debris extended out far enough to contact the plasmasphere of Earth. Once the two planets' plasmaspheres made contact, their electrical charge difference could act between them, creating a strong pull on the surface of the Earth. Initially, the Earth's crust lifted up in response, but once the plasmaspheres were in full engagement, an incredibly powerful thunderbolt of electrical discharge occurred, striking the Earth directly on the plain of Jordan and relieving the local electrostatic lifting force.

The resulting shockwave caused the highly stressed 65-million-year-old plate boundary fault to fracture, creating a colossal rift that extended for thousands of miles. Sodom and Gomorrah were destroyed by the thunderbolt, while the plain of Jordan was torn apart in a massive earthquake that caused huge areas of land to sink into the depths of the Earth. The shock was felt throughout the world.

Unlike the thunderbolt from Saturn that struck the Earth at the start of the Deluge, no rain fell from the sky, but on the island of Rhodes, snow-like flakes of gold from Jupiter's atmospheric tendril drifted down on the city. In most areas along the fault zone, fire and brimstone from the skies, and earthquakes and lava from the land destroyed almost all of civilization.

The Age of Jupiter (The Fifth Age)

Below are two quotes attesting to the birth of Venus at the start of the Age of Jupiter:

> The ancient Persians called Venus Tishtrya, "a magnificent and glorious star which Ahura Mazda [i.e., Jupiter] has established as master and overseer of all the stars." Plutarch described the events in the following terms: "Then Horomazes [Ahura Mazda], having magnified himself to three times his size, removed himself as far from the sun as the sun is distant from the earth...and one star, Seirios [i.e., Tishtrya, or Venus] he established above all others as a guardian and watcher." *(Jupiter, Gold, and the Birth of Athene)*

> The rain of gold on Rhodes is assigned by Pindar to the time when Athene was born from the head of Zeus. The expulsion of the protoplanet Venus from the body of Jupiter followed, by decades or by centuries, the contact of Saturn and Jupiter, and the people's interpretation regarded Venus as a child of Jupiter, conceived to him by Saturn. (ibid.)

The Age of the Dead Sea

Let's look at the geological features around the Dead Sea in the plain of Jordan that were created by Jupiter's encounter.

The rift in which the Lake of Galilee, the Jordan, and the Dead Sea lie is the deepest depression on any continent. The surface of the Dead Sea is close to 400 meters below the level of the Mediterranean, and its deepest bottom is some 320 meters lower still. The shore falls steeply from the Judean mountains on the west; on the eastern side of the rift rise the Moabite mountains. The walls of the chasm show sharp broken strata that remained horizontal, which proves that the breaking down was instantaneous. The force that caused this slide movement must have been stupendous. The ground of the rift around the Dead Sea is covered with coagulated lava masses, taking the form of an immense herd of giant elephants with rough skin. These lava eruptions from fissures are ascribed to the second interglacial period.

To the south end of the Dead Sea towers a big cliff of salt called Jebel Usdum (Mount of Sodom). According to *The Overthrow of the Cities of the Plain*, "It is absolutely impossible that the salt sediment of a sea should precipitate in such a form." The text further states, "Only the rupture of the ground could create this site, singular in the entire world."

These two Velikovsky quotes date the formation of the dramatic features of this area to significantly different time periods. The first, which is representative of the Conventional Wisdom's timeline, is "ascribed to the second interglacial period," which is said to have started around 250,000 years ago. The second is said to have taken place at the time of the Old Kingdom in Egypt, which is said to have ended about 4,200 years ago, a date that coincides with my unconventional wisdom's timeline.

River Flow

These dramatic geological changes forced the Jordan River to reverse its flow direction. Here are six quotes with comments on the water features of the area and the timing of their appearance:

> The story of the violent changes that occurred in the Jordan Valley, the memory of which is connected with the time of the patriarchs and in which Sodom and Gomorrah were overturned, does not mention that the Valley of Sittim, where the cities were located, became an inner sea. Sulfur and brimstone fell from heaven. One of the best-cultivated areas was overturned; fire from beneath and fire from above accomplished the desolation—all this is described, but not the appearance of a sea. However, when the Israelites under Moses and Joshua reached the area during their flight from Egypt, they found the lake there. It seems to have appeared after a catastrophe later than the one that destroyed Sodom and Gomorrah. *(The Great Rift and the Jordan)*

> The Great Rift, which begins in Syria between the Lebanon and Anti-Lebanon, runs along the Jordan Valley, the Dead Sea, the Arabah, the Gulf of Aqaba, the Red Sea, and continues through

The Age of Jupiter (The Fifth Age)

the continent of Africa as far as Zimbabwe. It is generally regarded as the product of a grandiose revolution in the shell of the Earth: for many thousands of kilometers, the Great Rift runs from Asia to Africa. (ibid.)

Prehistoric man witnessed the latest phases of widespread tectonic movements that convulsed East Africa and provoked great subsidences (sinkholes of as much as 1,500 feet or more) in the early Quaternary strata. These changes discharged lava and erupted scoriae, notably modifying the courses of rivers, the circumstances in which lakes rose or fell in level, and even changing the outlines of these bodies of water. (ibid.)

Whatever the structural changes of the Earth in the catastrophes before the one described here, there must have been some time when the Jordan streamed into the Valley of Sittim (the name of the plain before the Dead Sea originated) and continued into the Mediterranean, most probably through the Jezreel Valley. (ibid.)

Prior to the Exodus, the Jordan Valley was on a higher level than the Mediterranean Sea. With the rupture of the tectonic structure along the river and the dropping of the Dead Sea chasm, many brooks in southern Palestine that had been flowing to the south must have changed their direction and started to flow toward Palestine, emptying into the southern shore of the Dead Sea. This occurrence served as a symbolic picture for the dispersed Children of Israel, who also will return to their homeland: "Turn again our captivity as the streams in the south." (ibid.)

Now a simple reckoning shows that the saline sea with the Jordan has not existed longer than five thousand years. *(The Age of the Dead Sea)*

This last quote is based on the proportion of sodium to magnesium in the water of the Jordan as compared to that of the Dead Sea.

Total Destruction

The formation of the Great Rift caused extensive destruction over a wide area. The following seven quotes describe some of the devastation:

The opening of the Great Rift, or its further expansion, accompanied by the overturning of the plain and the origin of the Dead Sea, was a catastrophe that ended an era. In my understanding, the end of the Early Bronze Age or the Old Kingdom in Egypt coincided with these events. *(The Great Rift and the Jordan)*

The Old Kingdom in Egypt, the period when the pyramids were built—a great and splendid age—came to its end in a natural disaster. "At the conclusion of the Sixth Dynasty...Egypt is suddenly blotted out from our sight as if some great catastrophe had overwhelmed it." (ibid.)

In the same catastrophe, the civilizations of Mesopotamia and Cyprus were destroyed. What caused "the disappearance of so many cities and the upheaval of an entire civilization"? "It was an all-encompassing catastrophe. Ethnic migrations were, no doubt, the consequence of the manifestation of nature. The initial and real causes must be looked for in some cataclysm over which man had no control...Everywhere it was simultaneous and sudden." (ibid.)

In my scheme, the end of the Early Bronze Age or Old Kingdom in Egypt is the time of the momentous events connected with the story of the patriarch Abraham and described in the Book of Genesis as the overturning of the plain. The cause of the catastrophe could not have been entirely unknown to the ancients. We must also become attentive to other traditions connected with these events. (ibid.)

"The final end of the Early Bronze Age civilization came with catastrophic completeness...Jericho...was probably completely destroyed...Every town in Palestine that has so far been investigated

The Age of Jupiter (The Fifth Age)

shows the same break...All traces of the Early Bronze Age civilization disappeared." (ibid.)

"One of the most striking facts about the Early Bronze civilization is its destruction, one so violent that scarcely a vestige of it survived. We do not know when the event took place; we only know that there is not an Early Bronze Age city excavated or explored in all of Palestine which does not have a gap in its occupation between Early Bronze Age III and the Middle Bronze Age. To date this gap, we know that it must be approximately contemporary with a similar period in Egypt called the 'First Intermediate Period' between dynasties VI and XI (ca. 22nd and 21st centuries B.C.)." (ibid.)

The time of the patriarch Abraham witnessed unusual behavior by the planet Jupiter. The fact that Jupiter displayed a burst of activity exactly in the time of Abraham must not appear a coincidence: it was in the times of great global catastrophes, when the world was threatened with destruction, that religious reformers gained prominence, and contemporaries looked to a divine man for guidance. *(Zedek)*

Jupiter's Eastward Motion

During the time Velikovsky was researching "In the Beginning," he found references to an eastward motion of the planet Jupiter and struggled unsuccessfully to explain this improbable condition. He even briefly and somewhat grudgingly considered the possibility of Earth having been a satellite of Jupiter at that time as a possible explanation for the references. Below are two quotes on the subject that describe the essence of the problem:

> The Babylonians described Marduk, or Jupiter, as having an eastward motion, different from the other planets: "The earliest system from Babylon has, however, East and West reversed, and assigns to its chief god Marduk, as god of the planet Jupiter, a

definite easterly direction; to Mercury, on the other hand, a westerly one." *(The Change in Jupiter's Motion)*

The celestial mechanics of the implied reversal of Jupiter's apparent motion remains unsolved. Jupiter apparently changed the place of its rising points without a similar and simultaneous change by the Sun and all the planets and stars. It might seem that in order for Jupiter alone to be subject to a change, a reversal of orbital motion is required, an unlikely proposition from the point of view of celestial mechanics. *(ibid.)*

It is tempting to pursue the resolution of this apparent conflict the same way I did in Chapter 24, where I discussed a similar eastward motion reference for the Sun's movement immediately prior to the Deluge. However, since there was no "seven days of blinding light" for our current discussion, what I have considered to be the explanation for Saturn's participation in the Sun's apparent eastward motion does not apply in the current situation for Jupiter.

To fully understand the quote above, I have dissected the part that is in quotation marks phrase by phrase. There are three statements being made here. The first statement is: "The earliest system from Babylon has, however, East and West reversed," meaning that by today's directional conventions, westerly means easterly and vice versa. The second statement is: "… and assigns to its chief god Marduk, as god of the planet Jupiter, a definite easterly direction;" meaning that by Babylonian standards, Jupiter was moving in an easterly direction. Therefore, as discussed in the first statement, by today's directional conventions, Jupiter was actually moving in a westerly direction.

The third statement is "…to Mercury, on the other hand, a westerly one," meaning that by Babylonian standards, Mercury was moving in a westerly direction. Therefore, as discussed in the first statement, by today's directional conventions, Mercury was actually moving in an easterly direction. With these clarifications in mind, I will proceed to describe my interpretation of what the Babylonians actually saw.

The Age of Jupiter (The Fifth Age)

In the nighttime skies over Babylon, as over anywhere on Earth today, the planets of our solar system appear to drift slowly in an easterly direction against the background of the stars, except for the brief periods when the Sun, Earth, and the planet are close to being in alignment such that the planet under observation is high in the sky close to midnight, as was Jupiter at the time of its encounter with Earth. Under these circumstances, Jupiter's motion would appear to be retrograde, with its apparent motion in a westerly direction. Since Mercury lies between the Earth and the Sun, its retrograde motion is lost in the glare of the Sun. So when it is visible as a morning or evening star, it is always seen traveling in an easterly direction. With this phenomenon in mind, and in recognition of the stated fact that the Babylonians had the meaning of easterly and westerly reversed, the apparent message simply means that Jupiter was in retrograde at the time of its catastrophic interaction with the Earth 4,200 years ago.

Jupiter as God

Just as Saturn had been worshiped as a god after the Deluge, Jupiter's thunderbolts earned him a place among the gods, as the following two quotes imply:

> "It is conceivable that this planet was worshipped in that remote time and that, in the days of the patriarch Abraham, the cult of Jupiter was prominent in the Salem of the high priest Melchizedek. Melchizedek, 'priest of the most high' was, it follows, a worshipper of Jupiter. (Zedek)"

> "Marduk, the great god of the Babylonians, was the planet Jupiter; so was Amon of the Egyptians; Zeus of the Greeks was the same planet; Jupiter of the Romans, as the name shows, was again the same planet. Why was this planet chosen as the most exalted deity? In Greece, it was called 'all-highest, mighty Zeus'; in Rome, 'Jupiter Optimus, Maximus'; in Babylon, it was known as 'the greatest of the stars'; as Ahuramazda, it was called by Darius 'the greatest of the gods'; in India, Shiva was described as 'the great ruler' and considered the mightiest of all the gods; he was said

203

to be 'as brilliant as the sun.' Everywhere Jupiter was regarded as the greatest deity, greater than the sun, moon, and other planets." (The Worship of Jupiter)

Thunderbolts

Jupiter's primary cataclysmic action was to strike Earth with a "Fire from Heaven" in the form of interplanetary lightning called thunderbolts. This was not the first time thunderbolts had struck the Earth. Indeed, the Moon, Saturn, Phaeton, and Mercury all struck the Earth with thunderbolts as they came within Earth's plasmasphere. I have noted earlier that Phaeton even struck Mars with at least one thunderbolt as it passed by on its way toward Earth. In a future chapter, I will describe Venus striking the Earth and then repeatedly striking both Mars and the Moon with thunderbolts.

A thunderbolt is similar in composition and action to a lightning bolt from a passing thunderstorm here on Earth, but its magnitude is incalculably larger. A lightning bolt may destroy a house, whereas a thunderbolt can destroy a mountain range. Arguably, the Grand Canyon, Canyonlands, Needles Monument, and Bryce Canyon in the southwestern United States are the scars left by the impact of thunderbolts on the Earth. The Valles Marineris on Mars, a 4,000 km long canyon four times deeper than the Earth's Grand Canyon, is another candidate for the work of a thunderbolt.

In addition to the canyon-like scars, smaller short-duration thunderbolts can leave circular, crater-like scars on surfaces they contact. Telltale signs of this type of scar are their circular shape, a central peak of relatively undisturbed ground, and a single small-sized crater centered directly on the larger crater's rim. This characteristic signature is caused by the tendency of the negative charge on an electron to repel its neighboring electrons, thereby concentrating current flow at the outside diameter of a conductor, in this case, a column of plasma. As a result, when such a bolt hits a surface, the damage is concentrated in a ring with the center of the ring relatively undisturbed.

The Age of Jupiter (The Fifth Age)

The presence of a small crater on the rim results from the shutdown action of such a bolt, where the final weak discharge from the bolt looks for a high point on which to discharge its last trickle of energy.

One such incredibly large circular crater is Gale Crater on Mars, where "Curiosity," the Mars rover of NASA's Mars Science Mission, is actively exploring Mount Sharp, the crater's central peak, even as this book is being written. Photos of this mountain show sloping stratified layers of sedimentary rock, implying three conditions compatible with our discussions of this planet's history as a sister planet to Earth.

The first is the presence of sedimentary rock implicating the action of water-borne erosion debris, proving that Mars had water on its surface. The second is the slope of the stratification layers with respect to the local horizon, implying mountain-building activity commensurate with a shape change similar to that which I have discussed for Earth. The third is the undisturbed nature of the central peak, implying the action of a large-diameter electrical discharge bolt where the bulk of the current flow is around the periphery of the bolt.

The following seven quotes discuss the subject of thunderbolts at the time of the Jupiter encounter that started The Age of Jupiter:

"Nobody who observes a thunderstorm would arrive at the conclusion that the planet Jupiter sends the lightning. Therefore, it is singular that peoples of antiquity pictured the planet-god Jupiter as wielding a thunderbolt—this is equally true of the Roman Jupiter, the Greek Zeus, and the Babylonian Marduk." (The Worship of Jupiter)

"Pliny knew the origin of lightning in the friction of clouds—he wrote that 'by the dashing of two clouds, the lightning may flash out.' He did not confuse lightning with the thunderbolt that is discharged by the planets. He makes a distinction between 'earthly bolts, not from stars,' and 'the bolts from the stars.' Pliny knew that the Earth is one of the planets: 'Human beings are distributed all around the earth and stand with their feet pointing towards each other…Another marvel, that the earth herself hangs suspended and does not fall and carry us with it.' (ibid.)"

"In *Worlds in Collision*, the overpowering of one planet by another in conjunctions was quoted from the Hindu astronomical books; the electrical power which manifests itself in conjunctions is called "bala." Jupiter as the strongest planet is a "balin." (ibid.)

"We recognize in the change in Jupiter's motion the cause of great catastrophes in the solar system, which affected also the Earth in the age of the patriarchs, or at the close of the Old Kingdom. In that period Jupiter became the supreme deity, having removed Saturn from its orbit. Classical historians, speaking of the destruction of the Cities of the Plain, told of 'fire from the sky.' Tacitus narrated that the catastrophe of Sodom and Gomorrah was caused by a thunderbolt—the plain was 'consumed by lightning'—and he added: 'Personally I am quite prepared to grant that once-famous cities may have been burnt by fire from heaven.' Also, Josephus asserted that the cities had been 'consumed by thunderbolts.' Philo wrote that 'lightnings poured out of heaven,' destroying the cities." (Where a Planetary Bolt Struck the Ground)

"Since the time of Abraham was the period of Jupiter's domination that followed Saturn's and preceded that of Venus, we are led to the surmise that the thunderbolts which destroyed the plain with its cities originated from Jupiter, or from a magnetosphere or ionosphere overcharged by the nearby presence of the giant planet. Even today discharges leap between Jupiter and Io, one of its satellites. The charging of the Earth's atmosphere in the presence of Jupiter's huge magnetosphere prepared the way for a discharge: a planetary bolt struck the ground in the Valley of Sittim." (ibid.)

"The period was that of Jupiter's era of domination that followed that of Saturn and preceded that of Venus; and reference to the king and high priest Malki-Zedek ('My King is Zedek,' Zedek being

The Age of Jupiter (The Fifth Age)

the usual name of the planet Jupiter), in the days of the patriarch Abraham and of the destruction of Sodom and Gomorrah, seem to support my interpretation of the agent of the catastrophe. This very catastrophe caused the origin of the Dead Sea and also of the entire African Rift that extends from north of the River Jordan all the way through two-thirds of the length of Africa." (ibid.)

"Yet we are left with my original idea that goes back to the early forties—that the agent of the destruction was a bolt from Jupiter, or from the magnetosphere or ionosphere, overcharged by the nearby presence of the giant planet." (ibid.)

Chemical Deposits

Strong evidence exists of thunderbolts from passing planets as a source of chemical deposits in certain areas of the Earth, as testified to by the following eight quotes:

"The deposits of nitrate in Chile are found in a narrow strip over 1,400 miles in length, in the great desert in the northern part of the country. The origin of the nitrates is a problem that has not been solved." (The Origin of Nitrate Deposits)

"The Dead Sea, for many centuries proclaimed to be dead and capable of yielding nothing, is today one of the greatest reservoirs of natural nitrate under exploitation in the world, competing with the deposits of Chile." (ibid.)

"Nature is a great laboratory too. The Dead Sea region was the scene of an interplanetary electrical discharge when a powerful electrical spark leaped down from above or sprang up from the earth." (ibid.)

"A similar event created the Chilean deposits of nitrates, and the recollections of the Incas of Peru preserved the memory of this grandiose discharge. 'Fire came down from heaven and destroyed a great part of the people, while those who were taking to flight were turned into stones." (ibid.)

"It has been observed since ancient times that lightnings are attended by an odor of sulphur." (The Transmutation of Oxygen into Sulphur)

"Sulphur is one of the best insulators and static electricity, when accumulated on it, discharges in electrical sparks toward objects brought close to it. (ibid.)"

"Electrical discharges produced without the help of sulphur are also accompanied by the smell of it. This odor was referred to by Benjamin Franklin who, comparing lightning and electricity, wrote to the Royal Society in London that both phenomena are attended by a sulphurous smell. This he mentioned among twelve other properties which suggested that lightning is an electrical discharge. No importance was attributed by him or by anyone else since to this sulphurous smell. The smell of ozone is different from the smell of vaporized sulphur or sulphurous compounds, and the supposition that the ancients were unable to distinguish between the two disregards the fact that besides the smell of ozone a sulphurous smell follows an electric discharge". (ibid.)

"These stories of sulphur raining from the sky and the fearful expectations built upon them could be taken as fictions of an imaginative mind, were not the smell of sulphur an indication of its presence in the air following the passage of a discharge, and were not also the presence of sulphur deposits around the Dead Sea, thrust in deep below the ocean level, a substantiation of the story of the cataclysm." (ibid.)

Gold In Them Thar Hills

In 1848 gold was discovered at Sutter's Mill, California. It was mixed in with the sand at the bottom of the millstream and could be gathered by simply swirling water and sand around in a pan, allowing the heavier gold to separate. Once the word spread, a gold rush started that lasted until 1855. Then again, in 1898, gold dust was discovered mixed with sand in the Yukon Territory, setting off another gold rush. The Yukon gold

The Age of Jupiter (The Fifth Age)

was more difficult to recover since it was buried under permafrost, mud, and debris, but once exposed it could be separated with flowing water by panning or other separation methods. This was gold that had been deposited as dust and small nuggets close to the surface, not buried deep inside the earth. Similar deposits have been found in the Ural Mountains in western Russia and in other parts of the world. Jupiter brought this bonanza to Earth as the following three quotes reveal:

> "Pindar, speaking of the island of Rhodes, says that Zeus 'rained down on the city with golden flakes of snow' at the time Athene was born from Zeus' head, 'shouting with a far-ringing cry, and all Heaven and Mother Earth shuddered before her.' (1) Homer also says that 'upon them [the people of Rhodes] wondrous wealth was shed by the son of Cronus.' Strabo, after quoting Homer, adds that other writers 'say that gold rained on the island the time when Athena was born from the head of Zeus, as Pindar states.'" (Jupiter, Gold, and the Birth of Athene)

> "Rains of gold are reported also in the Chinese chronicles." (ibid.)

> "We conclude that the [Ural] chain became (chiefly) auriferous during the most recent disturbances by which it was affected, and that this took place when the highest peaks were thrown up, when the present watershed was established, and when the syenitic granite and other comparatively recent igneous rocks were erupted along its eastern edges." (ibid.)

The Age of Jupiter is the last age specifically discussed in Velikovsky's *"In The Beginning."* To understand the ages attributed to Venus and Mars, which follow in the next two chapters, I will use information from his published work *Worlds in Collision* as well as three quotes (presented in the final chapter of Part 4) containing references to Mars and Venus from the sections of Velikovsky's "In The Beginning" that discuss Saturn and Jupiter.

Five Friends

Chapter 27

Exodus & The Age of Venus (The Sixth Age)

Venus fell victim to a case of mistaken identity but was never the comet it resembled at one point. About 3,500 years ago, some 800 years after being ejected from its orbit around Jupiter in the days preceding the catastrophe of Sodom and Gomorrah, Venus could still be seen from Earth following its irregular orbit around the Sun. This orbit would bring it back to the vicinity of Jupiter's orbital path every few years. It had a tail of reddish debris acquired during its ejection from Jupiter, giving the planet the appearance of a comet. Since comets were considered a bad omen, succeeding prophets on Earth had been keeping a close watch on its position in the sky since its first appearance. On one of its passes near Jupiter's orbit, Venus had an encounter with Jupiter and was observed being turned in a direction that would take it on a course toward the Sun.

The prophets understood that this new trajectory would cause Venus to cross the orbital path of Earth, requiring close observation to determine its future course. As they watched and waited, their worst fears were confirmed when Venus swung around the Sun and was pulled into a trajectory that would take it on a collision course with Earth. With its tail energized by its proximity to the Sun, and with the Sun's backlighting giving the planet a crescent shape, Venus looked like the head of an approaching bull with the tail of a dragon streaming from its body—a truly frightening sight in the morning sky over Earth.

Every civilization on Earth, including Greek, Roman, Tibetan, Persian, Chinese, Incan, Aztec, Mayan, Hawaiian, Icelandic, and Israeli, has a story of Earth's encounter with Venus. One of the most dramatic and well-documented is the tale of the Israelites as told in the Bible. The tribe of Israel was being held captive in Egypt when Venus encountered Earth. The Bible tells of a series of "plagues" visited upon Earth as the encounter passed through its several phases. Let's review those plagues and their physical causes to promote an understanding of the essence of the Exodus (one of the Old Testament books of the Bible that notes these calamities).

As Venus approached Earth, its red tail was the first to make contact with Earth's atmosphere. As the encounter proceeded, a hot fine red dust rained down on all the inhabitants of the world. Rivers ran red. In Egypt, the Nile "turned to blood"—the first plague. The frogs that lived in the now-heated river climbed out onto the dry land and died, fish died, the water became putrid, and humans could not drink the water. This was the second plague.

The maggots that produce lice thrived on the dead frog carcasses, producing an explosion of lice—the third plague. Then thick clouds of flies appeared—the fourth plague. Anthrax ran wild through the Egyptian cattle, infected by the flies, resulting in the fifth plague.

After about two weeks, the red dust in Venus's tail changed its consistency to that of a fine ash from a furnace. It covered everything. The Egyptians, living in the city where the lice infestation was strongest, suffered from boils and blains (inflamed blisters) as the ash got into the sores from their lice bites—the sixth plague.

Four days later, as the Sun became dimmer and dimmer through the ash-filled skies, red-hot stones began to fall from the sky, accompanied by thunder and lightning no earthly storm could have produced. Petroleum, mixed with the falling stones, ignited and spread fire everywhere, even on the water, where the floating oil could not be extinguished. The Israelites had been warned by their elders that fiery hail was the next

Exodus & The Age of Venus (The Sixth Age)

likely component of the pestilence from the menacing horned-headed omen in the sky and poured river water on the thatched roofs of their huts to prevent them from catching fire. The effort paid off, saving their dwellings and cattle shelters, but nothing could be done to save the flax and barley crops that were ripe in the fields. The hail and fire from the sky was the seventh plague.

After the firestorm abated, the Israelites went out into the scorched fields and gleaned as much grain as they could, storing it away in sealed containers. The following day, a swarm of locusts rose up from the wetlands of the Red Sea in a swarm of epic proportions. The heat from the rain of fire caused the emergence of the entire locust population at once. An east wind blew the locusts over Egypt, where they descended on the scorched territory and devoured every grain of barley and every piece of fruit that had not been gleaned. The Israelites, with their sheltered cattle and stored grain, were prepared. The Egyptians' unsheltered cattle burned and with no gleaned food stored, faced starvation. This locust calamity was the eighth plague.

A few days later, the wind shifted and blew from the west, sending all the locusts back toward the Red Sea. But the darkness continued to deepen. For three days, the Sun could not be seen. Day was as black as night. Even the omen could not be seen. These three days of darkness constituted the ninth plague.

On the fourth day, the ground began to shake, and the sky lightened—not by sunlight, but by the light from Venus, the omen, which shone dimly through the pall as a full disk. As Venus moved closer to Earth, gravitational and electrical forces combined to distort Earth's crust, giving rise to intense earthquakes. The strong stone buildings of the Egyptians crumbled, killing a large portion of their population, while the reed and stick homes of the Israelites were spared since they flexed rather than fractured as the land on which they sat shook. These earthquakes, destroyed buildings, and loss of life were the tenth plague.

The distorted crust developed into a tidal bulge that locked Earth's crust into alignment with approaching Venus, stalling Earth's normal rotation. As Earth's crust stopped rotating, the atmosphere and seas continued rotating as before, creating powerful hurricanes that blew with ferocity for days in all parts of the world, along with tidal waves that surged onto the land. Parts of the world were in darkness for days. Other parts of the world saw the Sun in the sky for days on end.

At Venus's closest approach to Earth, its reflected light shone through the darkness, making it as bright as midday. The Egyptians, fearing more plagues, finally set the Israelites free, allowing them to flee toward the east until the same body of water and swamp that had created the plague of locusts blocked their passage. The Egyptian army, reneging on its promise, followed the Israelites and found them trapped between the Egyptians and the Red Sea, preparing to attack. At this opportune moment, Venus's movement carried it into a position where its electrical and gravitational pull drew the waters of the Red Sea to the north, revealing dry land and allowing the Israelites to cross. As the Egyptians tried to follow, a bolt of lightning between Venus and Earth discharged the electrical pull, and the waters rushed back into the seabed, drowning the Egyptian army.

Fifty-two years later, Venus made a second close pass by Earth. Again, Earth's rotation was stalled, but only for several hours this time. The abnormal movement of the Sun was noted by civilizations all over the world. Stones fell from the sky, the Earth quaked, and loud noises filled the sky. The walls of Jericho collapsed, but Joshua and his ram's horns took the credit. This was the last time Venus directly threatened Earth.

Chapter 28

The Age of Mars (The Seventh Age)

An observable spectacular cosmic dance was recorded by the ancients and later inspired Homer as he wrote the *Iliad*. About 2,700 years ago, Venus was still in an irregular orbit that occasionally crossed Earth's orbital path but had not directly threatened Earth for almost 750 years. Civilization had advanced to the point where accurate written records of celestial and geological events were recorded and preserved. King Tut had already been buried in the Valley of the Kings in Egypt. The nations of Rome and Greece had been founded.

One night, as people all over Earth were watching their local nighttime sky in the light of a full Moon, Venus could be seen approaching Mars much closer than usual. At their closest approach, the plasma spheres of the two planets touched, allowing a thunderbolt to flash between them. The discharge reduced their mutual electrical attraction, allowing the two planets to separate slightly. However, gravitational and residual electrical forces drew them back together again, resulting in a second thunderbolt discharge between them that equalized their electrical charge. As Venus disengaged, its new orbital path put it on a near-collision course with Earth's Moon.

As they approached, their plasma spheres (which were at different voltages) interacted, causing another huge thunderbolt to jump from Venus to the Moon, striking its Earth-facing side. Lava burst forth from the Moon and flowed out onto its surface in clear view of the awestruck spectators on Earth. After this encounter, with its electrical attraction neutralized, Venus shifted to an orbital path that never again threatened Earth. This amazing dance between Mars and Venus, along with a close encounter between Venus and the Moon, formed the basis for the battle of these gods as told in Homer's *Iliad*.

One of the Israeli prophets observed the new trajectories of both Venus and Mars after their epic battle in the sky and was able to predict that Venus would no longer be a threat to Earth. He also calculated that on its next two-year alignment (in 747 BC), Mars would pass very close to Earth, causing much the same devastation that Venus had caused during its encounter with Earth in the days of the Exodus. He warned his people of the coming catastrophe and was persecuted and eventually killed for his troubles. His predictions came true. Two years later, as Mars approached Earth, its gravitational and electrical pull distorted Earth's crust, stopping the apparent motion of the Sun in the sky and causing a massive earthquake that destroyed buildings and split open the mountains west of Jerusalem. The encounter changed the inclination of Earth's axis and its rotational period around the Sun.

Four more times, at fifteen-year intervals, Mars made passes very close to Earth, causing similar devastation to what occurred in 747 BC. The ancients were able to clearly see the two moons of Mars, Deimos and Phobos, during these encounters. In its final close encounter in 687 BC, showers of meteorites attended Mars's passing. After this final encounter, Mars's elliptical orbit became more circular, removing this last significant threat to Earth's security. To this day, Mars remains on a significantly elliptical orbit, bringing it close to Earth during its superior conjunctions.

The three quotes below, containing references to Mars and Venus, are from sections of Velikovsky's "In the Beginning" that discuss Saturn and Jupiter:

"In the *Iliad*, Ares-Mars is called 'fool.' Pallas Athena said to him: 'Fool, not even yet hast thou learned how much mightier than thou I avow me to be, that thou matchest thy strength with mine.'

The Age of Mars (The Seventh Age)

These words explain also why Mars was called fool: it clashed repeatedly with the planet-comet Venus, much more massive and stronger than itself. To the peoples of the world, this prolonged combat must have appeared either as a very valiant action on the part of Mars, not resting but coming up again and again to attack the stupendous Venus, or it must have appeared as a foolish action of going again and again against the stronger planet. Homer described the celestial battles as actions of foolishness on the part of Mars. Thus Kesil, or 'fool,' among the planets named in the Old Testament, is most probably Mars. (Khima)"

"The passage in the Book of Job (38:31) can now be read: 'Canst thou bind the bonds of Saturn and loosen the reins of Mars?' The bonds of Saturn can be seen even today with a small telescope. The reins of Kesil I discussed in *Worlds in Collision*, section 'The Steeds of Mars.' The two small moons of Mars, Phobos and Deimos, were known to Homer and are mentioned by Vergil. They were regarded by the peoples of antiquity as steeds yoked to Mars' chariot. (ibid)"

"At the end of the eighth century and the beginning of the seventh century before the present era, when every fifteen years Mars was approaching dangerously close to the Earth, Isaiah prophesied 'the day of the Lord's vengeance,' in which day 'the streams [of Idumea] shall be turned into pitch, and the dust thereof into brimstone, and the land thereof shall become burning pitch.' (The Transmutation of Oxygen into Sulphur)"

The Age of Mars and the transition from the Interim Earth to The Present Era, effectively ended with the orbital shift of Mars in 687 BC. After that time, no further catastrophes were brought about by planetary interactions with Earth. Of course, the people of Earth did not know it was over and continued to call it the Seventh Age or whatever number they happened to assign to the age where Mars had been threatening Earth.

However, for my purposes, The Present Era effectively began in 687 BC, or about 2,700 years ago. The stabilization of the chaotic solar system that resulted from the destruction of the flux tube at the end of the Interim Earth had been brought about primarily by the interaction of the plasma spheres of the planets. The space around the Sun, though swept clear of plasma by the shock wave from Geminga, still contained localized concentrations of ionized gas spilled from the ruptured flux tube immediately after the shock wave passed. These gases coalesced around the planets, forming teardrop-shaped plasma spheres extending hundreds of planetary diameters from the dark side of each planet.

As gravity-driven orbital mechanics took over from the flux tube's synchronous alignment forces, each planet, in turn, assumed its own independent orbital path, allowing it to be overtaken by the tail of planets orbiting closer to the Sun than itself. This interaction slowed the inner planet and accelerated the outer planet until all the planets were orbiting in such a way that they no longer intersected the plasma sphere of their neighbors. Even today, the plasma sphere of Venus extends a distance just shy of Earth's orbit, and the plasma sphere of Jupiter extends a distance just shy of Saturn's orbit, attesting to the stabilizing effect of an ionized gas tail that is the legacy of the long-gone flux tube.

All of life on Earth is fortunate indeed to have had a flux tube cocoon filled with plasma as an incubator of life during the Early Earth. That good fortune extended through the segue into the Interim Earth as Earth gradually changed shape without destroying the flux tube. It is entirely possible that the flux tube would not have formed in the first place under the conditions that existed around the Sun during the Interim Earth, in which case our Earth would never have existed at all.

Part 5

Calm and Understanding

Five Friends

Chapter 29

The Present-Day Solar System

Rather than fear Mars, today some of us imagine living there and raising a family. Ever since the short-lived Age of Mars effectively ended in 687 BC after Mars made its last threatening pass at Earth, the solar system has been calm and predictable. The Earth that civilization has inherited from the Age of Mars is the manifestation of the biblical Promised Land, wherein there are no more attacks from angry planet gods. We may still feel the occasional earthquake tremor, endure an isolated volcanic eruption, withstand a tsunami or two, and worry about the possible impact of a stray asteroid. However, we have not had to contend with the threat of an encounter with a planet-sized object from space, as was the case for every civilization on Earth since the destruction of the Golden Age of Saturn at the end of the Interim Earth.

We have remnants of the solar system's flux tube (ruptured 34,000 years ago) to thank for the present calm. The same plasma from which our Sun was born, and that filled the flux tube to protect Earth from the harsh environment of space as life evolved, has gradually been separating us from and circularizing the orbits of our planetary neighbors through the mechanism of plasmasphere interaction. This process, though not without occasional traumatic episodes, is clearly demonstrated by the seven Earth Ages presented in Part 4, Chapter 21.

Fortunately for us, the same plasma that controls our ultimate destiny still permeates the universe we inhabit today, providing a conductive path for the life-giving flow of electricity that powers every star in the cosmos. We do not have to fear that our Sun will run out of nuclear fuel and explode sometime in the future since it, like every other star, is merely one of an innumerable collection of lightning rods that glow at various intensities in response to intergalactic electrical currents flowing from one region to another.

If we must worry about things over which we have absolutely no control, then we should worry not about *whether* but *when* that galactic electrical current will either increase or decrease catastrophically.

The story I have presented shows that the galactic current can change gradually, as it did 65 million years ago when the Age of the Dinosaurs came to an end during the slow transition from the Early Earth to the Interim Earth. But that change was merely a slight dimming of the light.

If the local galactic current were to slowly decrease to a very low level, the arc-mode current flow at the Sun would first drop to the corona-glow mode and then to the dark-current mode. All heat and light from the Sun would cease, and Earth would freeze solid. Perhaps an industrious, cooperative society could burrow deep into Earth to use the heat of its core to survive for a while as our planet continued to orbit the darkened Sun. However, without the heat and light from the Sun, our civilization is doomed.

If our Sun's filament of galactic current were to increase dramatically, our Sun could experience a destructive stellar fissioning that would expose our planet to a burst of energy so powerful it would be burned to a cinder before subsequently being plunged into the absolute zero of deep space. Life would no longer exist anywhere near where our Sun once shone.

These doomsday scenarios are, fortunately, very remote. In the 3-billion-year time span covered in my presentation of our solar system's history, there has been only one influential galactic current change of consequence. The gradual decrease in current 65 million years ago at the end of the Early Earth took approximately 30,000 years to slowly fracture the Earth by changing its shape from highly ellipsoidal to more spherically ellipsoidal while increasing its effective surface gravity. Even if Earth had been spherical throughout the time of this slow-motion event, the Earth's crust would not have fractured, and its mammal population would still have survived the doubling of effective gravity.

Since Earth is now spherical and is orbiting the Sun under the influence of gravity alone, even if the event that ended the Early Earth were to happen again, major crustal fracturing would not occur. The only noticeable consequence would be a slight dimming of the Sun's light and heat output. Climate change would occur, but civilization would survive.

I also described how Earth changed its shape a second time when it became suddenly spherical as a result of the shock wave from the stellar fissioning of a nearby star that blew away our flux tube 34,000 years ago. That event ended the Transitional Earth and caused the traumatic re-fracturing of Earth's surface and the initiation of the planetary billiard game that sorted out the solar system until calm was finally restored about 2,600 years ago. If another nearby star experienced stellar fissioning, mass extinctions would again occur when the resulting radiation pulse reached Earth. However, with no flux tube to destroy, the debris wave that followed would have only minor effects since Earth is already spherical and orbits the Sun under the effects of gravity, which would not be altered by charged particle bombardment.

Throughout this book, I have systematically described an Unconventional Wisdom that provides credibility for the explanation of each particular phenomenon under discussion. Since critical parts of this Unconventional Wisdom depend on electricity, I have relied on the thorough presentation of the Electric Universe, or more formally Plasma Cosmology, contained in Donald E. Scott's book *The Electric Sky*. Dr. Scott includes many descriptions of telltale signs that attest to electrical activity presently acting in the solar system and in its recent past. However, he intentionally does not address the mechanism of the solar system's formation, nor does he discuss many of the geological features on Earth, Mars, and Venus that don't fit very well into the Conventional Wisdom but that do find a compellingly simple explanation in the Unconventional Wisdom presented herein.

As I have explored several subjects not discussed by Dr. Scott, it has been with the realization that the electrically powered Sun he so eloquently describes is an important part of my theory. Fortunately, the proof of the electrical nature of the cosmos is convincingly presented in *The Electric Sky*. Unfortunately, very little of that proof is found to be acceptable by proponents of Conventional Wisdom. If only Isaac Newton, the father of gravitational theory, and James Clerk Maxwell, the father of electromagnetic theory, had been alive a mere 2,700 years ago so that they both could have witnessed the battles that Venus fought with Mars and with Earth's Moon, what a difference there would be in the Conventional Wisdom being taught today. Hopefully, the detailed description of our solar system's formation and history as presented herein will add a new level of credibility to the theory of the Electric Universe.

The ongoing unprecedented era of planetary exploration, wherein the characteristics of the objects orbiting the Sun and those that orbit the large gas giants Jupiter, Saturn, Uranus, and Neptune, provide us with an ideal opportunity to verify critical features of our Unconventional Wisdom. It is imperative that this recently collected information be evaluated not simply in the light of the Conventional Wisdom's explanation that these objects are made up of billions-of-years-old galactic debris that could not quite form a planet and were brought to their current shape through the mechanism of co-accretion, but also from the perspective of the Unconventional Wisdom's tenet that the material from which these objects are made originated inside a flux tube.

Also proposed is that except for Earth, Mars, and the four gas giants, all are merely the randomly coalesced remnants of a planet I have called Proto-Venus, which fractured 34,000 years ago. The gradual accumulation of material inherent in the co-accretion process would have provided a much more uniform composition to these objects than the fractured-planet process would have provided, and all of the data thus far collected suggests that these objects are very different from one another.

The Present-Day Solar System

The Unconventional Wisdom that the solar system was formed within an electrically powered flux tube can find support in the recent discovery of a minor-planet-sized flux tube in operation within Jupiter's plasmasphere. Though this flux tube is dramatically smaller and structured differently than the proposed Sun-to-Proto-Saturn flux tube I have been describing, its mere existence lends credence to the proposition that a substantial flux tube can operate in the solar system environment. As a further benefit, its accessibility provides an ideal opportunity for the study of large-scale flux tube characteristics. As the largest remnant of Proto-Saturn, Jupiter retains the characteristic of having a voltage difference between its poles sufficient to form a flux tube by driving an electrical current from pole to pole. Jupiter's innermost moon, Io, which at 3,640 km is only slightly larger than Earth's Moon at 3,474 km, is enveloped inside the recently discovered flux tube, making it an ideal body around which an orbital probe could circle while measuring electrical and magnetic characteristics of the tube.

The study of the incredible diversity included in the composition of the solar system's asteroids, comets, small moons, large moons, minor planets, and even the planets Mercury and Venus is a subject suitable for numerous doctoral theses and is well beyond the scope of this book. Suffice it to say that every object in that broad list of solar system bodies can credibly trace its origin back to the breakup of Proto-Venus as it tried unsuccessfully to transition from a prolate spheroid to a sphere at the end of the Interim Solar System. The study of the non-planetary debris orbiting the Sun holds the key to verifying the validity of the whole Unconventional Wisdom theorem.

It is appropriate for me to review a few examples of the nature of this debris diversity to whet the appetite of those who may decide to perform more extensive studies of these objects. Consider NASA's Deep Impact mission that intercepted comet Tempel 1 on July 4, 2005, smashing a copper projectile into its surface and recording the impact with the fly-by portion of the spacecraft. Comets had always been thought of as dusty snowballs made up of icy cores with accumulated dust on their surfaces. The examination of the data returned from the mission showed that the debris cloud raised by the impact had less ice and more dust than expected and was much brighter than expected—so bright, in fact, that the view of the impact crater was obscured.

Six years later, when NASA succeeded in diverting a different comet-exploring spacecraft to approach Tempel 1, the photographs returned to Earth revealed a 150-meter-diameter crater at the projectile's impact point with an unexpected mound at its center. The Unconventional Wisdom we have been exploring expects comets to be more like rocky asteroids than snowballs and to be electrically charged to a voltage close to that of the solar plasma at their farthest distance from the Sun (aphelia). This voltage is significantly different from that of a fast-approaching spacecraft coming from a region of the solar plasma much closer to the Sun. As a result of these conditions, the expected encounter would involve a significant electrical discharge consisting of two flashes at the time of impact: a small flash as the projectile penetrated the comet's electrified plasmasphere, followed by an extremely energetic flash just before surface contact as a thunderbolt jumped the gap between the two objects to equalize the electrical charge. This is precisely what occurred. The resulting impact crater contained a raised central mound due to the tendency of the current flow in an electrical discharge to occur primarily on its outer surface rather than along its centerline.

Another interesting object is 67P/Churyumov–Gerasimenko, an irregularly shaped comet about 4 km wide at its widest point, recently visited by the Rosetta spacecraft launched by the European Space Agency (ESA). On November 12, 2014, Rosetta released its landing craft, *Philae*, which unintentionally bounced twice before becoming the first spacecraft to soft-land on a comet's surface. Unlike the high-velocity impact of Deep Impact, a copper projectile, the Philae landed without creating an electrical discharge crater because it had been orbiting the comet for four months prior to the landing, allowing the spacecraft and comet to reach electrical equilibrium. The lander detected organic molecules containing carbon, the basis of life on Earth, and reported a jagged, rock-like object with a hard surface beneath a 3- to 7-inch layer of dust, estimated to be as hard as ice. However, even when mission control activated a drill designed to cut into ice, the surface could not be penetrated, even at the drill's maximum setting. It is probably rock.

In 2011 and 2012, NASA's Dawn spacecraft visited and orbited Vesta, a 525-km-diameter heavily cratered rocky asteroid that is the second largest in the asteroid belt. It was then redirected to Ceres, an almost perfectly round, 950-km-diameter dwarf planet made of rock and ice that is the most massive object in the asteroid belt. Rocky Vesta has a surface pockmarked by many craters, none of which are large enough with respect to the asteroid's diameter to threaten its structural integrity if they were impact craters. Images of Ceres taken from the vicinity of Vesta initially led scientists to expect it had a smooth surface, indicating that resurfacing had occurred from the watery composition of its crust. However, when Dawn arrived at Ceres, the returning images showed a surface as rocky and cratered as Vesta's.

The two most famous asteroids surely must be Deimos and Phobos, the rocky moons that were captured along the way by our neighboring planet Mars. Deimos, at 14.5 km (9 miles) long, has a 3.2 -km (2-mile) diameter crater, and Phobos, at 27.4 km (17 miles) long, has a 9.7 km (6-mile) diameter crater. It is hard to imagine an impact with an object massive enough and traveling fast enough to have created craters this large on stony objects this small without shattering the objects themselves. Yet, there they are. The much more likely explanation is that these craters were the result of electrical thunderbolts as these asteroids passed close to other asteroids or planets with significantly different voltages. The Electrical Discharge Machining (EDM) process that results from such an electrical encounter vaporizes material with electrical discharge energy rather than impact energy, thereby leaving the cratered object intact and virtually unmoved.

At the other end of the size spectrum from comets and asteroids are the five particularly large moons in orbit around Saturn or Jupiter. Of particular interest is Titan, Saturn's largest moon, which, at 5,151 km in diameter, is bigger than Earth's Moon at 3,474 km in diameter and larger than Mercury at 4,878 km in diameter. Its orange-colored atmosphere is known to be primarily hydrocarbons and nitrogen. It has been shown to have lakes of methane and ethane on its surface in the polar regions, a solid normal ice surface floating on a liquid water ocean flooding a dense ice shell with a hydrous silicate core, and a highly eccentric orbit around Saturn.

The two questions we should ask about this intriguing object are: where did the hydrocarbons come from, and where did the water come from? The answer to both of these questions is "the flux tube."

Recall that back in Part 2, Chapter 10, a scenario is presented that allowed hydrocarbons to condense and fall as oil on the early Earth's surface long before there was life or even water on the planet. That oil sank into sediment deposited by the oil's runoff from higher elevations to lower basins and has become the shale oil and other oil deposits reached by deep drilling and even fracking in today's world. Keep in mind that the source of these hydrocarbons was the atmosphere of the flux tube that was shared by Earth, Mars, and Proto-Venus. In addition, recall that back in Part 4, Chapter 27, while describing the plagues that befell the Egyptians just prior to the Israelites' escape from captivity at the onset of the Age of Venus, I attributed the "hail and fire from the sky" that constituted the seventh plague to a petroleum rain. The area of Earth where this plague occurred is now referred to as the Middle East and is the source region for some of the most readily available crude oil on Earth today.

That oil also had its origins in the flux tube, but this time its immediate source was the remnants of the flux tube's original atmosphere that were captured by Jupiter and Saturn after the flux tube was destroyed at the end of the Interim Earth. Venus brought part of that atmosphere along after its ejection from Jupiter at the beginning of the Age of Jupiter. Saturn's Titan could easily have acquired its hydrocarbon atmosphere from that same source of flux tube atmospheric remnants.

The next issue I would like to discuss is the source of the water that makes up 50% of Titan's mass, as well as the four other large moons, all of which happen to orbit Jupiter. These are, in order of their proximity

The Present-Day Solar System

to Jupiter: Io, a rocky, slightly elliptical moon about 3,640 km in diameter with virtually no water; Europa, a rocky moon at 3,121 km in diameter that has a water-icy crust about 100 km thick; Ganymede, the solar system's largest moon at 5,262 km in diameter, which is about 50% water ice; and Callisto, another huge moon at 4,820 km in diameter, which is also about 50% water ice. The water contained on Saturn's Titan and three of the four moons of Jupiter accounts for about 96% of all the water in the solar system. By comparison, Earth holds only about 1.2% of the solar system's water.

It is highly probable that Proto-Venus did have an ocean of water on its surface before being shattered as the flux tube was destroyed. However, unless that ocean was more than 140 km deep—52 times the 2.7 -km average depth of Earth's oceans—all the water locked up in the outer solar system debris objects had to come from somewhere else. The first hint of where this water came from appeared back in Part 2, Chapter 10, and again in Chapter 11, as we discussed how the atmosphere of Early Earth was in equilibrium with the flux tube's atmosphere, such that excess water vapor in the flux tube would occasionally inject water vapor into Earth's atmosphere, causing flooding of low-lying continental shelf areas of Pangaea. A second hint appeared when we discussed Noah's flood in Part 4, Chapter 24, noting that some of the floodwaters fell from the sky as rain when a fragment of Saturn's atmosphere made contact with Earth. These two hints tell the story—namely, that the "somewhere else" was the remnants of the flux tube's atmosphere that congregated around Saturn and Jupiter.

As scientists collect data from existing and future spacecraft exploring comets, asteroids, Mercury, Venus, Pluto, the moons of the four gas giants, and even from our own Moon, it is very likely that, as in the examples discussed, the preponderance of evidence will support my Unconventional Wisdom's thesis: All this debris came from the fractured remains of Proto-Venus and the atmospheric gases from the flux tube. In fact, our Apollo astronauts, as they walked on the Moon's surface, have walked on Venus's surface material (now on our Moon) and have already brought samples of Venus's surface back to Earth.

Five Friends

Chapter 30

Climate Change

Climate change is nothing new. It has been happening on all three of the original planets in one form or another ever since they first formed. This book has described how Earth, Mars, and Proto-Venus were created inside a flux tube and, until recently, shared an interconnected atmosphere. When a climate-altering event, such as an explosive volcano or a lava flow, happened on any one of the planets, the attenuating effects of the flux tube volume acted to reduce the impact of that event on the particular planet where the event occurred. It is also important to note that the interconnecting volume would also cause these same reduced effects to be experienced somewhat later and to a somewhat lesser extent on the other two planets.

All that changed at the end of the Interim Solar System about 34,000 years ago. This interconnectivity between planetary atmospheres is worthy of its own Aha! moment:.

Aha! #17:
The three primordial planets shared a common atmosphere until 34,000 years ago. In today's world, our atmosphere is no longer connected to a flux tube and is not shared with anybody else. What happens on Earth stays on Earth. So, be careful out there and treat our atmosphere with respect.

The dramatic events at the end of the Interim Solar System saw the flux tube destroyed and its shared atmosphere dissipated. Mars lost its ocean and virtually all its atmosphere, Proto-Venus lost its crust and ocean and developed an entirely new toxic atmosphere, and all three planets were exposed to direct solar UV radiation for the first time. A special Aha! moment is set aside for this last revelation:

Aha! #18:
Earth, Mars, and Venus never saw direct sunlight or were exposed to UV radiation until 34,000 years ago.

Fortunately for the inhabitants of Earth, our planet was the only one to retain essentially the same atmospheric composition and pressure it had developed over its long life. There was sufficient oxygen, carbon dioxide, and water vapor to develop an ozone layer and to trap and hold enough solar energy to provide a temperate climate.

Since that fateful moment, and up until about 2,600 years ago, Earth saw remnants of the primordial atmosphere from the destroyed flux tube literally rain down on its surface in the form of floods, icy meteorites, and manna from heaven. But over the last two millennia, our planet's climate has finally stabilized, and its wide variety of flora and fauna is still thriving.

Now, as our modern world enters the third millennium AD, we find ourselves pondering the effects of observable global warming and climate change, which is causing ocean levels to rise, average ocean and air temperatures to increase, and more frequent severe weather conditions such as hurricanes, tornadoes, and blizzards to ravage our planet's surface. There is a spirited debate among scientists and politicians as to what mechanism is causing these effects and what can be done to control them to avert a potential disaster.

One of the methods proponents of Conventional Wisdom use to understand the causes of these effects is to look back at climate indicators

Climate Change

preserved in core samples from Antarctic and Greenland ice, ocean sediment, and terrestrial rock to understand past weather history. They often examine what are known as "Milankovitch cycles," which describe the collective effects caused by variations in Earth's movement through space, such as orbital eccentricity (413,000-year cycles), axial tilt (41,000-year cycles), and spin axis precession (25,771.5 -year cycles). These cycles influence variations in the solar radiation reaching Earth and its subsequent effect on climatic patterns.

The Unconventional Wisdom presented in this book contends that, since Earth and its sister planets were trapped inside a flux tube between our Sun and Proto-Saturn until 34,000 years ago and Earth has been in its present undisturbed orbit for only the last 2,600 years or so, Milankovitch cycle data dating from before 34,000 years ago are irrelevant. This precludes their use, or the use of any climate data prior to that date, as an argument for or against any particular climate change theory.

In further support of this argument, let me direct your attention to Chapter 20, "*The Storytellers.*" Of interest is a discussion of ice core data from the Greenland and Antarctic ice sheets and my ice sheet dating discussion from *The Cycle of Cosmic Catastrophes*. Also, consider my thoughts on the lack of rocky glacier-like moraines at the edge of the Antarctic ice sheet, my discussion of erratic boulders, drumlins, kettle ponds, and lateral and terminal moraines, and my musings on the round rocks of New England from *Cataclysm!*. My conclusion is this: There has only been one Ice Age on Earth, and it started 34,000 years ago. Before that time, there was no ice at all anywhere on Earth.

This is the heart of the cautionary tale I put forward here. Let the debate rage on as to the cause of global warming—be it increasing greenhouse gases, continued ozone layer depletion, or any other theory that might arise—but please use only data from the last 34,000 years or, better still, from the last 2,600 years.

This statement is important enough for a final Aha! moment:

Aha! #19:
Global warming theories should only use temperature data from the last 34,000 years during Earth's only Ice Age, or better still, from the last 2,600 years.

Below is a brief summary of the mechanisms involved in the two most prominent global warming arguments: the "Too little O_3" argument (insufficient ozone) and the "Too much CO_2" argument (excess greenhouse gases).

The "Too little O_3" argument starts with the Sun's radiation arriving at the top of the ozone layer in the form of visible light and three types of ultraviolet light: low-frequency UV-A radiation, medium-frequency UV-B radiation, and high-frequency UV-C radiation. Visible light, UV-A radiation, and a small portion of the UV-B radiation are not absorbed by the ozone layer and pass through to Earth's surface. UV-C radiation is completely absorbed by the ozone layer, where its high energy splits individual oxygen molecules into two oxygen atoms. These atoms then attach to two other oxygen molecules to form two ozone molecules. . .

Most UV-B radiation is absorbed by the ozone layer, where its somewhat lower energy level cannot split stable oxygen molecules but can split two unstable ozone molecules back into three oxygen molecules. By this process, the ozone layer continuously recycles oxygen into ozone and ozone back into oxygen in an eight-day cycle. If the ozone layer is depleted by chlorofluorocarbon gases (CFCs), such as spray-can propellants (which still persist in the upper atmosphere despite being banned in 1989), or volcanic eruptions that emit ozone-depleting chlorine and bromine, then more "hot" UV-B radiation reaches Earth's surface, exacerbating global warming.

The "Too much CO_2" argument hypothesizes that global warming occurs due to excess absorption of solar energy. This is because the two

Climate Change

most abundant greenhouse gases in Earth's atmosphere—water vapor and carbon dioxide—allow visible light and UV-A radiation to pass through the atmosphere but block the re-radiation of heat in the form of infrared radiation from Earth's surface back out into space. This process causes a net increase in atmospheric and surface (water and land) temperatures.

Recall that in Chapter 20, "The Storytellers," I discuss *In the Beginning*, noting the data sent back to Earth by Mariner II in 1962. That data indicated that the surface of Venus is very hot (800°F), thereby validating the youth of the planet and supporting one of the key revelations first published in Immanuel Velikovsky's *Worlds in Collision*. Shortly thereafter, Carl Sagan proposed that Venus's high surface temperature was due to what he called the Runaway Greenhouse Effect. Ever since then, scientists have used that example to project what global warming might do to Earth.

My theory for the high surface temperature of Venus is described thoroughly in Chapter 18, "Gone in a Flash," with a dramatically different explanation that refutes Sagan's theory. It does not preclude greenhouse gases as being the cause of Earth's global warming, but it should give pause to those who would use Venus as a vision of a future Earth. And, at the risk of being repetitive, there was no direct sunlight on Earth's surface capable of causing the greenhouse effect until after 34,000 years ago.

The determination of which of these two mechanisms, a combination of the two, or some other mechanism entirely is responsible for our present global warming episode is in the hands of scientists well-educated in Conventional Wisdom. If they use only data from the last 34,000 years—or, preferably, from the last 2,600 years—I will fully support their conclusions. However, I will remain very suspicious of any conclusion that does not.

Five Friends

Chapter 31

Closing Arguments

This chapter contains the closing arguments for the provocative rewrite of our solar system's history that I have presented in the preceding 30 chapters. Condensing this comprehensive thesis into a few pages cannot avoid losing the thread of logic that ties these chapters together. Those who have not been reading along up to this point are encouraged to do so now so that this summary makes sense. Here is the remarkable story of *Three Planets…* in three paragraphs:

Once upon a time, our Sun was the dominant star in a binary star system. Three rocky planets were aligned like pearls on a string within an electromagnetic flux tube that stretched between the Sun and its companion and was filled from end to end with a thin atmosphere.

The planets were egg-shaped, having been distorted from spherical by the same electrical forces that formed the flux tube and kept them in their synchronous orbits. The planet that was closest to the Sun was inhabited by carbon-based lifeforms, such as dinosaurs. Now called Earth, this planet retains evidence of the dinosaurs in the form of fossilized bones. The largest of their species were, at 100 tons, ten times larger than any land animal on Earth today. Fortunately, the same forces that distorted and aligned the three planets acted to offset the otherwise impossible weight of these animals, allowing them to walk around and dominate the planet.

About 65 million years ago, these electrical forces gradually subsided to about half their original strength over a time span of 30,000 years.

The resulting gradual loss of lift made it increasingly more difficult for successive generations of all land-animal species to stand, resulting in a gradual mass extinction where, by 64 million years ago, no land animal with an adult weight of more than 50 pounds was still in existence. The loss of lift also caused the three rocky planets to slowly become slightly more spherical, and in the process, their surfaces became distorted, creating rifts and raising rounded, Appalachian-like mountain ranges.

From the time of this shape-shift to about 34,000 years ago, the three planets and two stars were stable, allowing those inhabitants that had survived to adapt to the new effective gravity. Near the end of this period, humans evolved.

This Golden Age abruptly ended when a shockwave from the explosion of a nearby star swept through the solar system, taking less than an hour to destroy the flux tube and the electrical force field that had held the system together. With the loss of the flux tube, the common atmosphere shared by the three planets dissipated into space.

The Sun's companion star did not survive the shock, shattering into the four large gas giant planets now called Jupiter, Saturn, Uranus, and Neptune. With the electrical forces gone, gravity was finally able to collapse the three egg-shaped planets to near-perfect spheres, causing massive earthquakes, extinctions, and the rapid emergence of jagged Rockies-like mountain ranges. The first and second planets from the Sun, namely Earth and Mars, survived. The third planet did not; instead, its debris became Mercury, Venus, the asteroid belt, Pluto, and all the moons now in the solar system. By 687 BC, all the objects in our solar system had settled into the orderly arrangement seen today.

Yes, to most readers (scientist or not), this biography of the solar system sounds like pure fiction. After all, the scientific consensus is that the Earth is now and always has been spherical. Until my thesis evolved, no one had seriously proposed that the Earth was egg-shaped for most of its existence and has been spherical for only the last 34,000 years, and no one had imagined that ours is indeed a tale of two solar systems: one

Closing Arguments

with three planets and a companion star orbiting the Sun that lasted until 34,000 years ago, and the one we see today. I started out looking for an explanation of how dinosaurs were able to walk on Earth and wound up with a completely different explanation for how our solar system came to be arranged the way it is today.

In the pages of this book, I have presented a logical explanation that will convince the careful reader that this seemingly impossible story is, in fact, what really happened. In this endeavor, I have used only hard evidence: continental land masses, mountain ranges, sea floor cracks, coal deposits, fossils, sediment layers in rocks and ice, rounded rocks, and moraines. Myth, legend, lore, and ancient symbols are occasionally cited only for incidental local color and to reinforce otherwise deduced observations.

The first step of my investigation has been presented in Chapters 1, 2, and 3, where I describe the Dinosaur/Gravity problem in detail. Since the scientific community finds the size of Argentinosaurus to be non-problematic, and with a peer-reviewed paper to support their position, any argument to the contrary would not be immediately embraced by those supporting the Conventional Wisdom. With their predilection in mind, I started developing a means for providing enough lift so that the dinosaurs would feel an effective gravity of only one-third of Earth's present gravity, allowing them to be as mobile as modern-day elephants even though they weighed ten times as much.

The first approach I studied was whether gravity gradients and centrifugal force could offset 66% of Earth's gravity. Chapter 4 describes why that approach would not work and serendipitously provided my first of several *Aha!* moments: *Aha! #1: A spherical Earth and dinosaurs are incompatible.* It also hinted that if the primordial continent Pangaea was placed on one end of an egg-shaped (prolate spheroid) Earth, as discussed in Chapter 5, the stability issue of *Aha! #1* could be resolved.

My next step was to see if electricity, the only other force in the universe that can compete with gravity and centrifugal force, could give me the Earth shape I needed. The answer, as I described in Chapter 6, was: Yes, it can!! And the configuration that solved the problem also provided *Aha! #2: Different voltages in the wall of the flux tube and the DL point cause radial and axial forces capable of creating an egg-shaped planet.* Just the effect I was looking for.

By the time I got to Chapter 9, a complete description of my proposed Early Solar System had been developed, leading to *Aha! # 3: Electrostatic forces necessary to hold Proto-Saturn (the second star of the binary system) in orbit and to stretch Earth are compatible.* If this had not been the case, my quest would have ended right there, and this book would not have been written. Further pondering on this peculiar new solar system arrangement led to *Aha! # 4: Proximity to the gravity-only orbit radius (an orbital mechanics phenomenon) allowed only three planets.*

In Chapter 10, that same radius proved to be the inspiration for *Aha! # 5: Primordial Continent-Orientation (PCO) matters,* and was instrumental in other revelations that followed, including the formation of life on Earth. During my description of the evolution of Earth's flora and fauna, I pointed out that my timeline does not deviate from the last 600 million years of the Conventional Wisdom's timeline but that, on occasion, I do present differences of opinion as to the causation of certain features in that timeline.

Chapter 11 begins with a detailed description of Earth's geography just before the end of the Age of Dinosaurs, showing that the regions where Argentinosaurus fossils and coal beds are found today were all located close to the shore of Pangaea, adding credibility to *Aha! #6: The location of coal deposits is compatible with an almost hemispherical Pangaea at one pole of an elliptical Earth.* I continued by describing Earth's gradual shape-shift that caused the dinosaurs' extinction by the doubling of Earth's effective gravity over a period of 30,000 years, rather than an asteroid impact as the Conventional Wisdom proposes, and added *Aha! #7: The Chicxulub crater is consistent with a flux tube-to-Earth thunderbolt initiated by fracto-emission.*

My discussion in Chapter 12 presented a description of the breakup of Pangaea into six continents by forces much more powerful than those proposed by Conventional Wisdom. I further described how Conventional Wisdom's description of how Earth formed would have located Earth's ocean basin as a ring around the equator with a primordial continent at each pole—certainly not what we see today.

Chapter 13 followed with a detailed description of how Earth's gently rounded mountain ranges were formed, some in sinuous ridges, some in less well-defined wrinkles in the Earth's crust, and how extensive cracks opened up in the ocean floors, generally aligned perpendicular to the equator.

Life on Earth at the beginning of the Interim Earth is described in detail in Chapter 14. Only ten percent of the flora and fauna from the Age of the Dinosaurs and no land mammal with an adult weight greater than 50 pounds were in existence. A simple sketch showing the dramatic difference in the maximum size of land animals in the Early Earth, the Interim Earth, and Present Earth provided a graphic *Aha! #8: There is visual verification of effective gravity difference*, and irrefutable proof that the effective gravity on the surface of the Earth has increased over time as described in my theory.

In Chapter 15, the first hint of the impending breakup of the Interim Solar System was described as presenting itself in the form of a flash of light and lethal radiation from a supernova 40,000 years ago, an event that provided early humans with their first view of an object that was outside the solar system. The radiation from Geminga caused extinctions and mutations throughout the animal kingdom.

Humans that survived developed bigger brains and remembered the event by considering a bright light in the sky as a bad omen. I have described the images that these evolving humans saw in the sky overhead during the time period we now refer to as the *Golden Age of Saturn* in Chapter 16, including an important sketch of the three-planet/two-star solar system to show why they saw what they saw. Chapter 17 speculated as to how that view may have inspired their more intelligent brains to develop the concept of time.

Chapter 18 described the cataclysmic moment when the shockwave of high-energy protons and atomic nuclei from Geminga slammed into and destroyed the flux tube that held the Interim Solar System and its common atmosphere together. The resulting chaos that transformed the three-planet/two-star solar system into the Transitional Solar System is described in detail as all the actors in our Present Solar System began the sorting-out process. Key to that process was identifying the gravity and electrical charge of Saturn as the mechanism capable of capturing Proto-Venus and then lifting the debris of its outer layer from its core, in the process revealing *Aha! #9: Saturn lifted and redistributed the fractured crust of Proto-Venus.*

Another critical moment shortly after the shockwave impact is described as *Aha! #10: Gyroscopic precession lifted Earth's & Mars' axes*, revealing how Earth and Mars had their spin axes lifted up to their present-day 67-degree positions. The fact that Earth and Mars have almost identical spin rates—24 hours/day vs. 24.6 hours/day, respectively—and almost identical spin axis inclinations from the plane of the ecliptic—66.5° vs. 65°, respectively—whereas all the other present-day planets have widely divergent spin rates and axes tilt, provides strong confirmation of my thesis.

So far, I have focused my closing arguments on Earth-based evidence that supports my theory. But Earth was not alone during the Early and Interim Solar System. Mars and Proto-Venus were its companions, and they experienced the same historical events that shaped Earth. Chapter 19 presented descriptions of features on the surface of Mars and Venus and in the composition of meteorites that are observable today, lending evidence-based data to my basic thesis. Five Aha! moments—three for Mars, one for Venus, and one for meteorites —summarized those findings well. On Mars, the existence and size of a "V"-shaped feature in the boundary between the highlands and lowlands was presented as *Aha! #11: Mars mirrors Earth*. The presence of a "V"-shaped cut in the edge of an otherwise uninterrupted continental shape provided strong evidence that Mars also experienced the two shape-shifting events that fractured Earth. *Aha! #12: Sinkholes, but no continental drift on Mars* clarified how Mars accommodated those shape-shifts without fracturing into multiple continents as Earth did. *Aha! #13: Short-term heavy meteor showers on Mars around 34,000 years ago* used the pattern of impact craters on the surface of Mars to bracket the timing and duration of the meteor showers that caused them. Over on Venus, the observation that the present-day mean surface temperature of Venus, at 735 K, is a reasonable cooldown temperature to expect for a 1,900 K Venus-sized planetary surface exposed to deep space cooling for 34,000 years provided a strong inspiration for *Aha! #14: Present -day Venus' surface temperature and lack of impact craters are compatible with Proto-Venus having shed its original surface at the end of the Interim Earth* (not a runaway greenhouse effect).

The presence of iridium in meteorites was shown to be completely compatible with my theory's proposed source and led to *Aha! #15: Iridium in meteorites came from the shed shell of Proto-Venus.*

Chapter 20 presented one final Earthbound Aha! Moment: *Aha! #16: Crustal curvature change is observable in ancient dry lakebed shorelines.*

Chapters 20 through 28 offered observational evidence in the form of eyewitness accounts from our ancient and not-so-ancient ancestors that describe the Transitional Solar System as it sorted itself out from the cataclysmic events of 34,000 years ago. Since eyewitness accounts, even in present-day courtroom testimony, are notoriously unreliable, these observations were not used as the basis of my theory but serve merely to buttress its validity.

Chapter 29 presented some relevant closing thoughts and the description of corroborative data gleaned from news briefs contemporary with the researching of this book regarding several scientific probes to objects in our solar system. The descriptions of these discoveries were often preceded by phrases such as "Scientists are puzzled by the latest data from…," but when viewed in light of my thesis, the discoveries fit very nicely. The clincher will be when a mission to the asteroid belt succeeds in obtaining a core sample from an asteroid showing sedimentary rock—perhaps even with fossils—just what one would expect to find in a fragment of Proto-Venus's crust. Perhaps a review of samples brought back to Earth from the Apollo Moon landings will also reveal just such fragments.

Chapter 30 brings together climate change references from earlier chapters and discusses the effects of observable global warming and climate change on present-day Earth, which are currently being debated among scientists and politicians. The main actor in this discussion is the volume of the shared atmosphere in the flux tube. That volume is described as tending to dilute the effects of a climate-altering event on the planet where the event happened and substantially reducing the effects shared by the other two planets, thereby leading to Aha! #17: *The three primordial planets shared a common atmosphere until 34,000 years ago.*

Once the flux tube was destroyed, there was no more atmosphere sharing. Since then, what happens on Earth stays on Earth. The flux tube loss also prompted Aha! #18: *Earth, Mars, and Venus never saw direct sunlight or were exposed to UV radiation until 34,000 years ago.* References are made to climate information, including ice core data, being evaluated as scientists try to understand current climate change. The reminder that there was no ice on Earth before 34,000 years ago and that Earth has been in its present undisturbed orbit for only the last 2,600 years prompted the final Aha! moment: Aha! #19: *Global warming theories should only use data from the last 34,000 years, or better still, from the last 2,600 years.*

The chapter ends with the presentation of the "Too little O_3" argument (insufficient ozone) and the "Too much CO_2" argument (excess greenhouse gases) theories on global warming.

Chapter 32

A Brief Timeline of the World

The following timeline of the world differs dramatically from that of the Conventional Wisdom. It is strongly recommended that the detailed discussions presented in this book be thoroughly reviewed before reading this summary, thereby providing an understanding of the reasoning behind these differences.

THE EARLY SOLAR SYSTEM

At Least 3 Billion Years Ago—The Sun Forms

Electrical current flows within one of the spiral arms of the Milky Way galaxy in a flux tube filled with dust, hydrogen, and ionized gas. These materials cluster together at one of the Double Layer (DL) points and, through the z-pinch effect and mutual gravitational attraction, form a sphere about 860,000 miles in diameter. That sphere acts as a focal point that concentrates the local electrical field, creating a sustained arc that emits white-hot light. A star, our Sun, is born.

At Least 2.5 Billion Years Ago—A Companion Star and a Flux Tube Form

The electric current flowing in our local galactic flux tube increases, causing a portion of the surface of the Sun to be blown off, forming a second, much smaller star with a dramatically different voltage. This allows a flux tube to develop between the Sun and its new companion, Proto-Saturn.

About 1.5 Billion Years Ago—Three Planets Form

Three rocky planets —Earth, Mars, and Proto-Venus —form at DL points within the billion-year-old flux tube and orbit the Sun in synchronous rotation with Proto-Saturn. All three are prolate spheroids (ellipsoidal) in shape and spin with their long axes aligned with the line between the Sun and Proto-Saturn.

They each have variable effective surface gravity that is stronger near their equators and weaker near their poles. This effect is due to both the directional electrostatic forces that produced their ellipsoidal shape and the varying distance of these surface areas from the planet's center of gravity. The flux tube prevents gases from escaping into space, enveloping the three planets in a primordial atmosphere containing oxygen, nitrogen, and other more complex gases, including water vapor, some of which condenses, allowing 60% of the Earth to be covered with water, primarily on its Sun-facing end. The entire Earth is bathed in perpetual sunlight diffused by the gases in the flux tube.

Around 1 Billion Years Ago—Life Begins to Evolve

Life develops in the oceans of the early Earth while the flux tube acts to protect the new life forms by shielding them from external cosmic radiation. The charged atmosphere within the flux tube occasionally generates electrical discharges that create ionizing radiation strong enough to alter the DNA of life forms, causing mutations that can lead to either the further evolution or extinction of a species.

In the last 100 million years of this time period, dinosaurs evolve to be the dominant land species, with some growing to a length of 100 feet and weighing 100 tons—a size that would not allow them to stand up in the one-G world we live on today.

A Brief Timeline of the World

THE INTERIM SOLAR SYSTEM

Around 65 Million Years Ago—The Age of the Dinosaurs Ends

The twin star system with its interconnecting flux tube drifts into a part of the galaxy where the plasma electric charge is weaker, allowing the two stars to gradually spread farther apart over a period of 30,000 years. All three planets in the flux tube experience reduced electrostatic forces such that they can no longer sustain their initial ellipsoidal shapes. The areas adjacent to their equators expand, and their pole separation contracts as they slowly become less ellipsoidal and more spherical while their effective surface gravity increases.

On Earth, the resulting crustal fracturing and folding create rift valleys and raise chains of mountains. Lava belches forth, volcanoes erupt, and smoke and ash fill the sky, causing one of the largest extinctions life on Earth has experienced. Up to 90% of all species are annihilated. Only those species with an adult weight of 50 pounds or less survive. Virtually all the dinosaurs become extinct, allowing small mammals, which are less affected by the increased gravity, to survive and become the dominant species on the planet. Some small dinosaurs survive to eventually morph into birds. The single landmass facing away from the Sun begins to fracture and break apart as the poles become closer to spherical in curvature. Plate tectonics begin. Due to its much smaller size, Mars is much less affected by its changing shape, but Proto-Venus barely survives while experiencing many of the effects seen on Earth.

About 40,000 Years Ago—The Golden Age of Saturn Begins

Mammals have evolved to be the dominant species on Earth. The radiation pulse from a supernova that had exploded 300 years earlier in the constellation Gemini reaches Earth, killing most of the species on the planet. The few early humans that survive the radiation pulse evolve with a greater intellectual capacity due to radiation-induced gene mutation, allowing them to tell stories and record their history. They worship the sight of the companion star that had only recently become visible through the thinning flux tube atmosphere and is now constantly overhead. Every tribe on Earth sees it as their own God, giving it its own special name, each of which translates today as Saturn. These ancients are the biblical "Adam and Eve" and are the humans back to whom we can all trace our DNA. They live in a temperate climate bathed in the perpetual diffuse sunlight of an eternal spring with abundant food for the taking—a true "Garden of Eden."

THE TRANSITIONAL SOLAR SYSTEM

About 34,000 Years Ago—The Age of Uranus Begins

The charged particle wavefront from the supernova of 6,300 years earlier slams into the solar system, destroying the flux tube and causing the companion star Proto-Saturn to break apart into four large gassy planets: Jupiter, Saturn, Uranus, and Neptune, in a nova-like explosion. The loss of the flux tube electrostatic forces causes all three rocky planets to suddenly change their shape from a mild ellipsoid to a sphere. On Earth, the shock of this change creates cataclysmic crustal upheavals and mass extinctions, but Earth survives. Mars survives a similar experience. Proto-Venus, however, does not survive its shape change. Instead, its crust and a portion of its upper mantle buckle and fracture. It is quickly captured by Saturn and settles into a highly elliptical orbit.

With each low-altitude pass, gravity gradients and electrical charge differences lift the loose pieces of fractured overburden from the new surface of Proto-Venus into separate orbits around Saturn, where the larger pieces are soon shaped into spherical bodies by their own individual gravity. This leaves behind its core as one Earth-sized planet (Venus) and a field of large pieces, which soon form two smaller-than-Mars -sized planets (Mercury and Phaeton) and numerous spherical moon-like objects of various size and composition. All these objects are pulled into erratic orbits around the four new gassy giants, while the smaller pieces of leftover debris spread out, forming the asteroid belt that now orbits the Sun.

Earth, Mars, and the newly formed gassy giants with their newly acquired satellites in tow assume erratic solar orbits, while some of the remaining debris smashes into the surfaces of all the scattered objects in a chaotic transitional solar system. The Sun appears as a small bright disk in the sky, smaller than their God Saturn appeared to them before the cataclysm. With the flux tube gone, its diffuse atmosphere is also gone, allowing people on Earth their first view of a starlit nighttime sky.

They can also see, in order of brightness, Saturn, Jupiter, Uranus, and Mars. Having moved to a remote orbit, Neptune is not immediately noticed. The loss of the flux tube atmosphere also allows direct sunlight to shine on the Earth's surface at full intensity for the first time in Earth's history. Areas near the equator are initially heated to new high temperatures, while the poles are exposed to a nighttime sky at a temperature near absolute zero.

Ice forms on Earth for the first time in its history, as the continent of Antarctica develops a year-round ice surface, and the ocean water at the North Pole forms extensive fields of sea ice. By 26,500 years ago, the only ice age Earth has ever seen is covering much of North America, northern Europe, and Asia, profoundly affecting Earth's climate by causing drought, desertification, and a large drop in sea levels. A warming trend started in the Northern Hemisphere approximately 20,000 years ago and in Antarctica at approximately 14,500 years ago, consistent with evidence for an abrupt rise in sea level at the same time. This now global warming trend is increasingly manifesting itself as climate change in all parts of the world.

About 20,000 Years Ago—The Age of the Moon Begins

As Uranus and Saturn approach each other on one of their passes around the Sun, the stronger gravity of Saturn pulls Uranus into a slingshot maneuver, sending it into a much larger orbit. In the process, the largest of Uranus's captured moons is stripped away and sent on a near collision course with Earth, where it is captured to become Earth's own permanent Moon.

It settles into a near-circular orbit much closer to the Earth than its present distance. The gravitational and electrical interaction during the approach and capture lifts mountains to new heights, raises immense ocean tidal waves, and fractures the Earth, allowing lava to spill forth. Humans fear and respect this new God in their heaven. It is their supreme deity as the largest and brightest object in their sky.

About 11,500 Years Ago—The Age of Saturn Begins

A second shock wave from the supernova arrives at the newly formed chaotic solar system, disturbing the orbits of Saturn and Jupiter so that they pass in close proximity to each other. As they approach, Jupiter's mass strips away two of Saturn's satellites the size of small planets, sending them off into space. At its closest approach, Jupiter's gravity attracts a portion of the electrically charged atmosphere of Saturn and tears it away, causing a nova-like explosion.

A segment of the charged atmosphere is thrust toward the Earth's plasmasphere, and as it makes contact, strong electrical forces stress the mantle of the Earth, fracturing its surface, causing volcanoes to erupt, and dramatically lifting mountain ranges to new heights. Some of the rocky and icy pieces of debris from Saturn's atmosphere and the larger of Saturn's errant satellites approach the north pole of the Earth, where the debris enters the Earth's atmosphere as meteoric objects that smash into the surface of the Earth, sending fractured shards in a secondary impact pattern over a wide area. The errant satellite then passes directly over the north pole, where its gravity and electrical forces first lift the Earth's crust and ocean waters to heights that cover mountains and then, as it flies by, releases the water to effect a massive torrential flood causing extensive extinctions.

The close encounter with Earth changes the satellite's path, causing it to fall into the Sun, where it is destroyed. The Earth's bulging crust is shifted to a new position on the planet's molten core, while the core itself does not alter its spin axis. The crust's movement positions a previously temperate region 1,000 miles farther north, bringing on sudden local

A Brief Timeline of the World

cooling that freezes some mammals where they stand as the debris wave from the flood envelops them. After the flood, Saturn is no longer in its prescribed place in the heavens, and Jupiter is much more prominent. When Saturn is eventually detected, it appears much smaller, is moving more slowly across the sky, and is surrounded by a series of rings. The Moon has also been moved to a larger orbit around the Earth, changing its orbital period and making it appear smaller.

About 10,000 Years Ago—The Age of Mercury Begins

The smaller planet-sized satellite that was torn away from Saturn at the beginning of that planet's Age has been watched by the people of Babylon as a rogue star on a path that may encounter the Earth. They build a tower in the middle of the plain as a refuge lest another flood befall them. As predicted, this rogue star that they call Mercury approaches the Earth. It gets close enough that its plasmasphere makes contact with that of Earth, resulting in violent electrical discharges between the two bodies. One strong bolt hits the top of the tower as though it were a lightning rod. The bolt destroys the tower and subjects the people of Babylon to strong electrical currents that cause momentary amnesia and temporary confusion of speech...

The atmospheric disturbances cause violent hurricane winds to blow, destroying other structures, but it does not bring a flood as had been forewarned. Mercury changes its path during the encounter to an orbit that keeps it between Earth and the Sun, where it can thereafter be seen as a morning or evening star.

About 5,000 Years Ago—The Age of Jupiter Begins

Jupiter is the dominant star in the sky, holding Venus as a satellite in an unstable orbit. On one close encounter between the two, an electrical discharge disturbs Venus's orbit so that it escapes into an elliptical orbit around the Sun and, at the same time, releases a highly charged tendril of primordial debris toward the Earth. When this finger of plasma reaches Earth's plasmasphere, the voltage difference between Jupiter and Earth acts to pull the two planets toward each other, lifting a portion of the

Earth's crust in the process. As the mutual attraction increases, a powerful electrical discharge releases the mounded Earth, causing a colossal rift that extends for thousands of miles. The cities of Sodom and Gomorrah are destroyed by the thunderbolt and the entire Syrian plain sinks. The shock is felt throughout the world.

3,500 Years Ago—The Age of Venus Begins

On one of Venus's close approaches to Jupiter's orbit, the two planets interact again, sending Venus on a path around the Sun that intersects Earth's orbit. As Venus rounds the Sun with its reddish comet-like tail leading the way, it begins a complicated encounter with Earth. At its closest approach, Venus's gravity and electrostatic forces pull the Red Sea to the north enough to allow the biblical crossing of Moses and the Israelites. Then, with the discharge of the electrical forces in a thunderbolt, the water is released, drowning the pursuing Egyptian army. Fifty-two years later, Venus makes a second close pass with Earth.

2,700 Years Ago—The Age of Mars Begins

While the Moon is high in the Earth's nighttime sky, Venus makes an approach to Mars that is close enough to allow their plasmaspheres to make contact, with the result that a thunderbolt jumps between them. They separate and are then drawn together again, allowing another spark to jump between them. Venus is slowed by this second encounter with Mars and changes its orbit to move closer to the Earth and the Moon. As Venus passes close to the Moon, their plasmaspheres interact, allowing a huge electrical spark to jump between them as well. Lava bursts forth from the Moon and flows out onto its surface. These encounters will be recorded as the battle of these gods in Homer's *Iliad*.

After these encounters, Venus moves away from the Earth and never again crosses Earth's orbital path. Mars, on the other hand, changes its orbit so that in the year 747 BC, it passes very close to Earth, causing much the same devastation that Venus caused on its encounter with Earth in the days of the Exodus. To the west of Jerusalem, a massive earthquake destroys buildings and splits mountains asunder. This encounter changes the inclination of the Earth's axis, momentarily stops the motion of the Sun in the sky, and disturbs the Earth's rotational period around the Sun. Four more times, at fifteen-year intervals, Mars makes additional passes very close to Earth, causing much the same results as it did in 747 BC. In its final encounter in 687 BC, showers of meteorites attend its passing. After its final encounter, the elliptical orbit of Mars is shifted so that it never threatens Earth again.

THE PRESENT SOLAR SYSTEM

2,640 Years Ago—The Present Age Begins

Our solar system has settled into its present quiet form while global warming on Earth continues.

PS: The poles are running out of ice!

POSTSCRIPT

Researchers poring through data from NASA's retired Kepler Space Telescope have recently discovered a binary star system that they have named Kepler-47. This system has three planets orbiting around two stars that eclipse each other every 7.45 days, making it a circumbinary system. If Proto-Saturn had not broken apart into four large gassy planets—Jupiter, Saturn, Uranus, and Neptune —in a nova-like explosion 34,000 years ago, there would have been no Age of Uranus, and our solar system would have been dramatically different.

Five Friends

EPILOGUE

The five friends met for their Egg-Shaped Earth Society Peer Review meeting, this time for the purpose of discussing the theory proposed in the manuscript each had read and digested.

The astrophysicist was the first to offer his opinion: "I've read the manuscript and it's given me a lot to think about. I agree that for centuries the role of electricity has been given short shrift in the formation of our solar system. Gravity and heat at the surface of the Sun make sense once you add to the picture the electrical charges and the flux tubes they create. Even the elusive 'dark matter' is no longer necessary."

The climatologist chimed in, saying that his visits to Antarctica and the land masses bordering on the polar ice cap have convinced him that we can buy time by allowing development of these areas for human habitation while we study our options for inhabiting other earth-like bodies in our solar system.

The geophysicist added, "I found the manuscript convincing. I am thinking that in the short term, we can devise habitability techniques that will allow the population of Earth's polar regions while we study Mars and the moons of our outer planets to choose which of these planet-like bodies are appropriate destinations for future colonization."

The paleontologist added her opinion by saying that the habitat and diet of flora and fauna on Earth give her confidence that Mars and some of the larger moons of Jupiter and Saturn (at least!) may be suitable for colonization.

The engineer added, "Now all my profession needs to do is develop the means to get us settled somewhere else in our solar system before all the ice on Earth is gone."

Closing Remarks

The facts have been there all along,

Interpretation got it wrong.

My story redefines our space,

Friends and readers, I rest my case.

Respectfully submitted,

W. Clark Dean

ACKNOWLEDGMENTS

I am deeply grateful to all those who have contributed to the creation of this work.

To the wisdom of my parents Gardner and Effie Dean, who treated their four children to many summers of tent trailer camping as we explored 49 of the 50 contiguous United States. We got to see and hike mountains, valleys, caverns, shorelines, deserts, lava flows, and all the wonders of nature. Some of the explanations offered by park rangers and guides raised more questions than they answered and started me thinking at an early age about how the world we live in actually evolved to its present form.

To Rafael Sharon (psychoanalyst, and Immanuel Velikovsky's grandson), who gave me permission to quote Velikovsky and cite his works both published and unpublished. He also recommended "a few folks I should talk to." They Are: C J Ransom, who knew Immanuel Velikovsky and his theories; Laird Scranton, who wrote *The Velikovsky Heresies* about recent findings and how they validate Velikovsky; and James Kenyon who is part of the Thunderbolts Project dealing with the electric universe. C J, Laird, and James, thank you all for your help early on in this project when I needed it the most.

To Donald E. Scott; Thank you for writing *The Electric Sky*. Its revelations about the interior of the Sun and the convincing arguments you make about the electric nature of the entire universe resolves so many of its enigmas.

To Robert Kreitler, a life long friend who listened to my early concerns about dinosaurs and the formation of the earth, asked intelligent questions, and told me "You ought to write a book," so I did.

To Dr. Peter L. Ward, a climate change expert who wrote a book on global warming and attended the International Paris Conference on the subject. Thanks for reviewing my early manuscripts and providing rebuttals to my thesis backed up by recent Ice age data that you say proves your point, but I know it actually confirms my basic "Only One Ice Age" thesis. We can agree to disagree and still remain friends.

To Barbara Dee for her superb editorial skills and for guiding me through the process of publishing this book.

To Kristen and David, my two children for their unflagging support and up-to-date expertise in all things electronic making cell phones and lap tops almost user friendly to me. Their opinions on my writing at many critical decision points were always helpful.

My wife Lynne, the light of my life, never failed to support me during my professional career and throughout my retirement as I tilted with windmills on the way to making this book happen. She passed away peacefully on our 61st wedding anniversary less than a year before this book's publication. Without her dedication to our family this book would not be here now.

Any errors you may find as you read my story are completely my responsibility and mine alone. I hope you enjoy your journey through time as I believe it really happened.

ABOUT THE AUTHOR

Clark Dean graduated from Lehigh in Pennsylvania in 1963 with a B.S. in Mechanical Engineering. That year, he won the National ASME Old Guard competition for a technical paper and presentation on an internal combustion engine he invented in his senior project, and was awarded the "Andrew W. Knecht Award for the Practical Application of Engineering Theory."

Dean earned an M.S. in Mechanical Engineering from the Hartford Graduate Center of Rensselaer Polytechnic Institute (1968) and has had a diverse career that exemplifies "the practical application of engineering theory" in the automotive, aerospace, railroad, and computer printer industries.

His inventions throughout his career have earned some 30 U.S. patents. In addition, he received two NASA Tech Brief awards. His experience in engineering and physics, coupled with intense interest in astronomy, led to him building his own telescope and, later, designing critical parts of space suits and life support systems for the Lunar Module, the Space Shuttle, and the International Space Station. He also read volumes of scientific literature on the origins of space and our own Earth's birth and earliest eras. He authored *Five Friends Discuss Global Warming* as a direct result of his curiosity, problem-solving, and theorizing about the universe.

In the foothills of the Berkshire Mountains in Connecticut, Dean enjoys his retirement by staying active, hiking to nearby waterfalls, and canoeing on local rivers. In fact, his latest patented invention is a new design for a canoe paddle. He continues to write and future publications include an exploration of off-Earth human civilization options.

www.ingramcontent.com/pod-product-compliance
Lightning Source LLC
Chambersburg PA
CBHW041235060526
44107CB00136BA/747